Total Quality Service

Also available from ASQC Quality Press

Baldrige Award Winning Quality: How to Interpret the Malcolm Baldrige Award Criteria, Fifth Edition
Mark Graham Brown

Quality Management Benchmark Assessment, Second Edition
J. P. Russell

The Reward and Recognition Process in Total Quality Management
Stephen B. Knouse

Customer Retention: An Integrated Process for Keeping Your Best Customers
Michael W. Lowenstein

Measuring Customer Satisfaction: Development and Use of Questionnaires
Bob E. Hayes

To request a complimentary catalog of publications, call 800-248-1946

Total Quality Service

A Simplified Approach to Using the Baldrige Award Criteria

Sheila Kessler

ASQC Quality Press
Milwaukee, Wisconsin

Total Quality Service: A Simplified Approach to Using the Baldrige Award Criteria
Sheila Kessler

Library of Congress Cataloging-in-Publication Data

Kessler, Sheila.
    Total quality service: a simplified approach to using the
Baldrige Award criteria / Sheila Kessler.
      p.   cm.
    Includes bibliographical references and index.
    ISBN 0-87389-336-0
    1. Total quality management—United States.   2. Malcolm Baldrige
National Quality Award.   I. Title.
HD62.15.K47   1995
658.8'12—dc20   95-24290
                                            CIP

10 9 8 7 6 5 4 3 2 1

ISBN 0-87389-336-0

Acquisitions Editor: Susan Westergard
Project Editor: Jeanne W. Bohn

ASQC Mission: To facilitate continuous improvement and increase customer satisfaction by identifying, communicating, and promoting the use of quality principles, concepts, and technologies; and thereby be recognized throughout the world as the leading authority on, and champion for, quality.

Attention: Schools and Corporations
ASQC Quality Press books, audiotapes, videotapes, and software are available at quantity discounts with bulk purchases for business, educational, or instructional use. For information, please contact ASQC Quality Press at 800-248-1946, or write to ASQC Quality Press, P.O. Box 3005, Milwaukee, WI 53201-3005.

For a free copy of the ASQC Quality Press Publications Catalog, including ASQC membership information, call 800-248-1946.

Printed in the United States of America

 Printed on acid-free recycled paper

 ASQC
Quality Press
611 East Wisconsin Avenue
Milwaukee, Wisconsin 53202

Contents

Part 1   Introduction   **1**
Definition of Total Quality Service   3
The Baldrige Template for Success   6
How to Use This Book   7
How the Baldrige Award Criteria Are Incorporated   7
ISO 9000 Standards   9
Lovers and Criminals in Quality   10
Requirements for Successful Implementation   12
Schedule of Implementation   14
Summary   14
Notes   15

Part 2   What to Do   **17**
Introduction   19
Total Quality Service Road Map   19
What-to-Do Outline   21
Summary   26

Part 3   How to Do It   **27**
Introduction   29

   1.0   Leadership   **31**
   Introduction   31
   Total Quality Service Culture   33
   Champions   33
   Steering Committee   34
   Project Plan for Quality   36
   Declaration of Commitment   36
   Success Story System   39
   A Visibility Plan for Results   40
   Celebration System Tips   42

Quality Accountability System   45
Annual Quality Report   47

2.0   Information and Analysis Tools   49
Introduction   49
Criteria for Data Collection   49
Types of Information and Analysis Tools   51
Control Charts   52
Fishbone Diagrams   53
Check Sheets   55
Pareto Charts   56

3.0   Customer Focus and Satisfaction   59
Why Measure Customer Needs and Satisfaction?   59
Customer Needs Assessment   59
Customer Focus and Satisfaction Measures   60
Customer Satisfaction Surveys   61
Focus Groups and Interviews   70
Other Customer Satisfaction Measures   74
Internal Customer Satisfaction Measures   76
Benchmark Tips   77

4.0   Strategic Planning   81
Strategic Analysis   83
State-of-the-Organization Report   86
Industry Trend Analysis   89
Competitor Profile   89
Strategic Synthesis   91
Quality Analysis   91
Key Processes   93
Key Quality Objectives   96
Strategic Plan   97
Quality Control Plan   98
Procedures Manual   100

5.0   Human Resource Development and Management   103
HR Planning and Management   104
Employee Involvement Process   104
Suggestion System   107
Quality Improvement Teams   110
Employee Education and Training Plan   114
Employee Performance and Recognition   115
Measures of Employee Well-Being and Satisfaction   117

6.0   Process Management   **119**
   New (or Improved) Product (or Service)
      Design System   119
   Quality Function Deployment   121
   Production and Delivery Quality
      Management Processes   121
   Support Services Quality System   123
   Supplier Quality Management System   124
   Quality Assessment System   125

7.0   Business Results   **129**
   Summary   131
   Notes   131

Part 4   Resources   **135**

   **Bibliography and Books That Highlight
      Service Quality**   **137**

   Glossary   **143**

Index   **151**

Part 1

Introduction

Frontline employees not only *deliver* the goods in service organizations, they *are* the goods. Many service companies in the United States have stalled somewhere after getting an overview in total quality service (TQS). Service quality *sounds* good, but how do you do it—especially in the intangible service business?

Definition of Total Quality Service

The goal of service quality is very simple—customer satisfaction. High levels of repeat customers lead to high levels of profit. The Pareto rule applies here. Many companies find 20 percent of their customers provide over 80 percent of their revenue. These are repeat customers.

Service quality focuses on satisfying your customers' needs in the hundreds of little "moments of truth" that make up the customers' perception of you. Jan Carlzon, CEO of Scandinavian Airlines System, first put these moments of truth into focus. He described those moments when the customer comes eye to eye with your organization, through billing, advertising, telephone calls, reservations, complaints, and use of the service or product. Each moment provides a window into the rest of the organization. Customers generalize about the entire organization based on that one moment of truth.[1]

Satisfying these moments of truth, one at a time, means customers come back. Customer retention is at the heart of profitable companies. Frederick F. Reichheld, founder and director of Bain & Company's customer retention practice, looks to a better measure of quality: From his work with regional banks, Reichheld figures the 20-year customer is worth 85 percent more in profits than the 10-year customer. Reichheld has also found that boosting customer retention 2 percent can have the same effect on profits as cutting costs by 10 percent. A similar study revealed that a 5 percent increase in retention leads to significant boosts in profits, as shown in Exhibit 1.1.[2]

The sales cost of recruiting a new customer is five to ten times that of retaining a repeat customer.[3] Thus, with higher satisfaction you achieve a higher return rate. Your cost of sales go down and your profits go up. It's that simple.

If customer service is the main goal, what are the components of TQS? Noriaki Kano, a Japanese Union of Scientists and Engineers (JUSE) counselor, and department head at the Science University of Tokyo, is one of the leading experts on quality in Japan.[4] I went on an Executive Mission to Japan during September and October of 1992. In a lecture there, Kano said quality must be

- Scientific

- Systematic

- Teamwide

Scientific means that managers use data rather than their gut-level intuition to make complex decisions. Managers use data on quality—not slogans and exhortations—to drive decisions. Managers survey customers on their needs and levels of satisfaction. Managers do not assume they know their customers. Managers and teams set service

Industry	Boost in profits
Automotive service	81%
Auto/home insurance	84%
Bank branch deposits	85%
Professional publishing	85%
Credit cards	89%
Life insurance	90%
Corporate insurance brokers	93%
Advertising agencies	95%

Exhibit 1.1 Impact of retention on profits: What a 5 percent increase in customer retention does to profits.

quality goals, monitor them, and use the results to feed into the planning process. Results, rather than grandstanding, are rewarded.

Systematic means just that. Customer surveys are sent out regularly rather than when someone gets around to it. Systematic questions are included to determine trends. Feedback is used to set service quality goals and create innovations to meet needs. Solutions are monitored to see whether they are working. Critical systems that support quality, such as hiring, training, recognition, promotions, and information, are described later.

Teamwide means that everyone is on board. Service quality stretches from the top of the organization to the bottom. It stretches through the various departments of the organization as well. A survey of 450 organizations found that a majority of firms used quality practices in their manufacturing areas but not in administration, procurement, materials, and other support groups.[5] My own recent training programs, which include individuals from over 200 of the Fortune 1000 companies, reveal that this is slowly changing. One 1986 study found that over 75 percent of manufactured goods customers who left their suppliers do so because of the quality of the service, *not* because of the quality of the product.[6]

The intent of this book is to provide a step-by-step approach to service quality implementation. Many employees have complained that they see many books on quality, Deming, Juran, and theory, but few references on how to go about implementation. This book greatly simplifies the process of implementation in hopes that interested managers see what is needed in an easy-to-understand approach.

Not all of the steps need to be done at once. You can select which initial steps are practical and expand over years to include the Baldrige Award–level quality embedded in this guidebook. Do not be discouraged if you are doing very few of the steps in your current practices. From my experience the normal implementation takes anywhere from two to five years. AT&T Consumer Communications Systems, GTE Directories, and Wainwright, Baldrige Award winners in 1994, all started their journeys before 1987 and still "feel like they have a long way to go."[7] The Japanese took over 20 years to perfect the process.[8]

The first question many people ask is, How large does a company have to be before this much work pays off? The answer is one person. Plans are shorter and so is the coordination process with small organizations. The second question is usually, How much money does it cost? If done correctly, it will save the company money. The outlay is mainly in the quality of thinking, not in the volume of paper. Keep your reports short—10 pages maximum. Appendices and resources can be longer, but your main reports should be appealing and understandable to everyone in your group or organization. Zytec, a company of 650 people spent $9000, plus 767 hours of labor, to win the Baldrige Award.[9] The third question is, Can this be done with just one department within an organization? Yes. If the rest of the organization operates out of sync with your quality objectives, ask to be left alone to pilot quality practices and demonstrate success.

The Baldrige Template for Success

The step-by-step approach is based primarily on successful companies that have either won the Baldrige Award or have come very close. My 14 years of experience within two Fortune 100 companies and as a consultant and Baldrige Award examiner has brought me in contact with over 50 U.S. companies with excellent quality programs and eight Deming Prize-winning Japanese companies. The Baldrige Award is the Olympic gold medal for companies that are implementing quality. The Deming Prize is the Japanese equivalent.

The Baldrige Award criteria provide an excellent template for success. Several researchers have found a strong relationship between Baldrige Award winners and substantial increases in market share, profit, employee and customer satisfaction, and revenue. The General Accounting Office did a study on 22 Baldrige Award finalists and found a significant relationship between improved profits, customer service, employee satisfaction, and revenue gain and Baldrige Award scores.[10] Over 1000 companies in Japan have been practicing total quality for over 40 years.[11] The fact that Japan now owns 40 percent of the world's assets should say something. The Japanese own 27 percent of the world's top 100 banks. The United States owns two. Fifteen years ago, the United States owned nine of the ten world's largest banks.[12] Need I say more?

Some common denominators of implementation have been incorporated in this guidebook; for example, the depth of commitment to quality, the visibility of results, the use of data, and the use of celebrations to fan the fires of quality. None of this can be done without human spirit. Total quality service works best in a positive, fun climate. The Baldrige Award criteria doesn't spell out this critical dimension, but you can't get results without it. From 1990 to 1994, Southwest Airlines has been more profitable than all the rest of the major U.S. air carriers combined, because its managers respect the importance of human spirit.

According to Herb Kelleher, CEO of Southwest Airlines, the most important criterion for hiring is the personality of the applicant. "Hire the personality and train the skills" is his byline. "We look for a spirit." Thus, if you really want success, go back to basics in quality. The 3-Rs are recruiting, rewarding, and recognition. The first exercise in delegation is to give everyone a celebration budget. Then let different areas come up with their own ways of giving recognition and rewards.

How to Use This Book

This book is divided into three main parts: (1) Introduction; (2) What to Do; and (3) How to Do It. The what-to-do part walks you through starting up a quality effort in a logical outline form. Some of the steps involve plans, processes, and tools that may be new to you. The idea was to stay simple and not combine chronological steps and detail. The how-to-do-it part fully defines these new words. For example, you will find that you need to do a **strategic plan** and **customer satisfaction surveys** in the chronology of implementing the program. Note the bold letters. Everything in bold has a tip sheet in the how-to-do-it part. The emphasis in this book is on less words and more action. Too many companies have huge volumes of strategic plans that are underimplemented. The moral of this story is: ***Condense the verbiage and expand the behavior.***

This book is meant to invite creative thinking, not restrict it. You will probably find that the forms need to be modified to fit your specific needs. In reality, no guidebook can replace your creative energy while working through your implementation process. This book is intended to spark ideas, not limit them. A certain amount of chaos in TQS is a healthy sign. The enthusiasm by which these steps are implemented is much more critical than the step itself.

How the Baldrige Award Criteria Are Incorporated

This guidebook follows the Baldrige Award criteria, of which there are seven categories.

1.0 Leadership

2.0 Information and Analysis

3.0 Customer Focus and Satisfaction

4.0 Strategic Planning

5.0 Human Resource Development and Management

6.0 Process Management

7.0 Business Results

Only customer satisfaction was taken out of order from the actual Baldrige Award criteria. The reason for this change was because companies have achieved more financial success if they based their key quality improvements on what was important to their customers. I asked managers of eight Fortune 100 companies to list how their customers would answer the question, "What are the most important attributes of that product or service?" Most of the managers felt that they truly knew their customers because they had so much contact. Yet when I asked their customers the same question, the lists generated by the customers overlapped those of the managers from only 10 percent to 60 percent.

GTE Directories, winner of the 1994 Baldrige Award, was a prime example. GTE did considerable research on its customers during the late 1980s, but asked questions that the GTE managers deemed important to customers. As a result, GTE spent $15 million on a new press in Los Angeles to improve the technical quality of Yellow Pages ads. In 1992, GTE Directories started to ask the customers about what they considered important.[13] Customers listed the following elements.

1. Demonstration of value for Yellow Pages investment
2. Strong business relationship
3. Outstanding customer service
4. Cost-effectiveness

Understanding what was important to customers would have changed the priorities of many projects—if only GTE had asked the right questions in the 1980s. Thus, putting **customer focus and satisfaction** close to the beginning of the process is important.

Within each of the seven criteria, the Baldrige Award examination process looks for *approach*, *deployment*, and *results*. These terms are defined as follows:

- Approach: The way you go about meeting the criteria
- Deployment: How you implement what you plan
- Results: What results you get

Approach	• Uses appropriate and effective tools, techniques, and methods
	• Is systematic, integrated, and consistently applied
	• Uses evaluation and improvement cycles
	• Is based on data and information that are objective and reliable
	• Uses objective, qualitative, and reliable data
	• Is innovative
Deployment	• Applies to all requirements
	• Applies to all appropriate internal processes, activities, and employees
	• Applies to all appropriate product and service characteristics
Results	• Show current performance levels
	• Demonstrate results of quality goals and efforts
	• Are sustained through trends
	• Are compared with world and industry leaders

Source: 1995 Baldrige Award Criteria (Washington, D.C.: National Institute of Standards and Technology), 40.

Exhibit 1.2 Baldrige Award evaluation dimensions.

Companies score low on Baldrige Award audits if they are anecdotal, inconsistent, and fail to integrate their goals, monitoring devices, and results. Companies score high if they are goal-oriented, driven by customer preferences, and deploy their objectives throughout their ranks. Specifically, approach, deployment, and results get high marks according to Exhibit 1.2.

ISO 9000 Standards

What about the ISO 9000 series standards? ISO is the International Organization for Standardization, and ISO 9000 is a series of standards now being required of many U.S. companies. The European Union has established directives, which include the ISO family of standards, as methods of satisfying requirements to sell to European companies.

The ISO 9000 series is only one small part of a successful quality template. The ISO 9000 series promotes consistency. The main ingredients are having a **quality control plan** and having procedures that are consistently followed. Do not think that ISO 9000 has anything to do with a template for business success. It doesn't. You could be registered to the ISO 9000 standards and go broke. You could be making buggy whips with perfect consistency. Later revisions in the ISO 9000 series may address strategic planning, employee involvement, or customer satisfaction measures. For now, the Baldrige Award criteria does a better job in these areas. The discipline and clarity of job descriptions behind ISO 9000 standards are, however, healthy attributes to incorporate. This guidebook integrates some of the ISO 9000 practices in the section on process management. Special attention, however, will be required to get your company registered to ISO 9000.

Lovers and Criminals in Quality

One of the assumptions that is hidden between the lines of this book is that improvement needs to be both top-down and bottom-up. A common mistake for U.S. companies has been to jump into employee empowerment and then beat a hasty retreat when the bottom line begins to suffer.

Kano, the Japanese quality leader who has taken over much of the late Ishikawa's role in leading the quality effort in Japan, characterizes two types of quality improvement efforts.[14] One Kano calls "looking for criminals" and the other is "looking for lovers." Exhibit 1.3 shows how each process entails a different approach.

Notice that looking for criminals is best done by people who have hands-on experience with the likely culprits. Looking for lovers involves outsiders, open minds, and the willingness to try new ideas. This openness is imperative because many new product ideas come from customers!

Empowerment assumes a well-trained workforce. Trained in what? In job skills, in English as a second language, in math, in business issues, in team problem solving, and in quality. Unfortunately, the Americans are behind the Asians because the U.S. educational system is subpar. While the literacy rate in Japan is 99 percent, U.S. companies are spending 43 percent of their training budget to cope with illiteracy.[15]

Looking for criminals	Looking for lovers
Looks for inefficiencies in processes Involves detective work Has to be specific and thorough Wants to get rid of culprits Usually bottom-up	Looks for innovative ways to improve products/services Involves creative side Has to be open to ideas Has to nurture new ideas Looks outside of organization Usually driven by top of organization

Exhibit 1.3 Types of quality improvement efforts.

Just in the past few years have we begun to focus on positioning the U.S. educational system for the twenty-first century. Thus, implementation of a truly beneficial quality effort requires considerable direction from the top and gradual empowerment from the bottom.

Empowerment can happen immediately on issues that relate to each employee's job. Employees usually know best how to do their jobs. Baldrige Award–winning companies like Ritz-Carlton and Zytec make generous allowances for employees to spend money on fixing problems that relate to key customer satisfaction items. Ritz-Carlton gives each employee a $2000[16] spending limit; Zytec gives each employee $1000.[17] Federal Express will allow an employee to make what would be radical decisions, such as renting a helicopter, if the need is demonstrated, as shown in Tom Peters' film *Forgotten Customer.*

An employee **suggestion system** is an excellent example of bottom-up quality. It needs to be matched, however, with an equally rigorous top-down quality direction. That comes through the **strategic plan,** the mission statement, and **key quality indicators.** The direction also happens through systems that are supported by top managers in resources and time away from the actual job. The benchmarked Japanese Deming Prize winners spend anywhere from 10 percent to 20 percent of their working life in training and improvement projects.[18] The excuse, "We're too busy because we have so many crises" is opposite to the top-down management philosophy that will drive improvement. The Baldrige Award criteria also looks for this top-down integration through **key quality indicators,** the **strategic plan,** hiring practices, motivational systems, and training.

Requirements for Successful Implementation

In order for this book to work, you need

- A willingness to change
- A group that has a basic level of honesty and integrity
- A personal commitment to quality from several key people
- A culture that supports quality (positive rather than blaming)

Quality usually starts with just a few people in any organization. The higher those people are in the organization, the more likely the process is to succeed. Any single group can declare itself a quality group and thrive without the overall organization being on board—if it is allowed to operate independently. Plenty of examples of these highly successful units are seen in large U.S. companies, such as Saturn within General Motors and AT&T Universal Card Services within AT&T. In a *Quality Progress* article entitled "To Boldly Go Where so Many Have Gone Before," Karen Bemowski tells how Saturn developed is own unique labor agreements, organizational structure, and marketing operations within General Motors.[19] Paul Kahn, the CEO of AT&T Universal Card Services, formulated the company with the Baldrige Award criteria in mind. To help facilitate success, at least a few top managers of the group or company must be on board. Exhibit 1.4 allows you to do a quick assessment of readiness to implement quality efforts.

Count the number of checks in this list. If you had more than six check marks, you might try the following steps before implementing the processes in the rest of this book. AT&T Consumer Communications Services, winner of the 1994 Baldrige Award, chose all of these options as it implemented its continuous improvement program.[20]

1. Provide coaching for upper-level managers.
2. Have upper-level managers and the board of directors go through a seminar on service quality.
3. Do an upside-down review. Have subordinates rate the managers on quality readiness.
4. Have upper-level managers go through an executive retreat to assess this cultural readiness and work through a plan to change the culture. That is done by defining the desired cul-

❏	1	Managers talk as if the organization is strong on quality but they do not "walk the talk."
❏	2	There are too many chiefs and not enough workers.
❏	3	Managers refuse to hear data when it conflicts with their own opinion.
❏	4	Managers think they can delegate quality to others in the organization.
❏	5	Managers throw money at the latest fad in training and consultants without truly understanding ingredients to success.
❏	6	Managers use reorganization as their key tool to improvement.
❏	7	Managers' decisions are based on intuition rather than data.
❏	8	Managers don't listen to employees' suggestions.
❏	9	Subordinates are not consulted when promotions are considered.
❏	10	Managers spend more time on mistakes than successes.

Exhibit 1.4 Checklist for quality readiness.

ture and then doing a *gap analysis* by surveying employees to see how close the current culture is to the desired culture. Outside facilitators can be helpful.

Leadership training comes in many forms. There are four basic types: feedback, personal growth, skill building, and conceptual.[21] Each has its drawbacks and advantages. The latest trend is to use a value-based approach. Value-based leadership maintains that if we share certain values, the bond between us will be stronger than if we follow the same commands. Robert Greenleaf, an AT&T executive, detailed this approach in his book *Servant Leadership*. Southwest Airlines uses virtual leadership. It requires leaders to articulate the company vision and then create an environment where employees can figure out the answers.

Southwest started this type of leadership class for all employees in 1986 and now requires all senior executives to attend classes four times a year. Unlike many of the more academic approaches to teaching leadership, Southwest's program is 90 percent participatory and exercise-based. Its style has paid off in 1990s profits.

Levi Strauss has also employed virtual leadership training. At the Blue Ridge Plant in Georgia the old value of "check your brain at the door and sew pockets" has been traded for a worker-run plant. Workers are cross-trained for 36 tasks instead of just one or two. The Levi policy manual has gone from 700 pages to 50. As a result, flawed jeans have been reduced by 33 percent; the cycle time between an order and shipment has been reduced by 10 days; and the manufacturing time for a pair of jeans is one-fifth what it used to be. Levi Strauss has had five years of record profits under the virtual leadership principles of CEO Robert Haas. Controls are conceptual, not procedural.[22]

The Japanese went through 20 years of changing a very authoritarian management philosophy into one that involves considerable direction from the top and input and decision making at the bottom.[23] Total quality service is both top-down and bottom-up management. This book will provide a guide for doing both.

Caveat: Total quality service and/or the Baldrige Award template does not work in a group where the leadership is highly unresponsive. Work on leadership first if this is the case. Unresponsive leadership will only make employees and customers resentful as heightened expectations are dashed against the rocks of arrogance and resistance. Most organizations have a bit of this in place, but ultimately leadership has to walk the talk. As was cited in a *Fortune* article, "Ninety-five percent of American managers today say the right thing. Five percent actually do it." That's got to change.[24]

Schedule of Implementation

Every company has a different implementation schedule and pattern. Exhibit 1.5 illustrates how a sample schedule may look. It shows only the first year because most of the planning and systems are established then. The results from the first year are given to the **steering committee** so the process can be improved. Small companies can move faster than large companies that need more time. Very conservative companies take the longest time of all.

Summary

The purpose of Part 1 was to set the groundwork for the implementation of total quality service. TQS was defined as customer satisfaction.

First Year

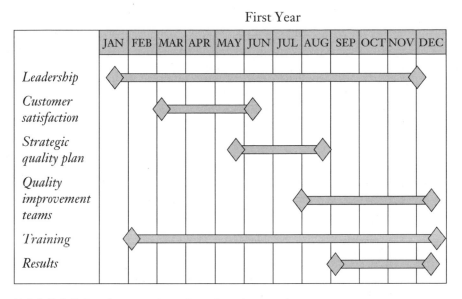

	JAN	FEB	MAR	APR	MAY	JUN	JUL	AUG	SEP	OCT	NOV	DEC
Leadership												
Customer satisfaction												
Strategic quality plan												
Quality improvement teams												
Training												
Results												

Exhibit 1.5 Implementation of total quality service.

To be successful, the TQS effort must be systematic, scientific, and teamwide. After the use of this book was explained, the Baldrige Award criteria and ISO 9000 standards were introduced. The importance of cultural readiness was emphasized and a quick measure of readiness provided. The last section of Part 1 highlighted the length of time it takes to implement service quality and provided a sample first-year schedule.

Part 2: What to Do quickly walks you through some of the implementation steps. These will be explained in Part 3: How to Do It.

Notes

1. Ron Zemke with Dick Schaaf, *Service Edge: 101 Companies That Profit from Customer Care* (New York: Penguin Books, 1995), 19.

2. Larry Armstrong and William C. Symonds, "Beyond 'May I Help You,'" *Business Week*, 25 October 1991, 101–103.

3. Madhav N. Sinha, "Winning Back Angry Customers," *Quality Progress* (November 1993): 53.

4. Jeremy Main, *Quality Wars: The Triumphs and Defeats of American Business* (New York: Free Press), 205.

5. "U.S. Firms Don't Recognize Role of Logistics Process," *Quality Progress* (March 1992): 14.

6. Consumer Complaint Handling in America: An Update Study (Technical Assistance Research Program Institute, Washington, D.C., March 31, 1986, photocopy).

7. George Lieb and Joe Nacchio, "Strategic Quality Planning" (paper presented at the Quest for Excellence Conference, Washington, D.C., 6 February 1995).

8. Noriaki Kano, interview by author, Tokyo, Japan, 28 September 1992.

9. Doug Tersteeg, Zytec quality manager, Redwood Falls, Minnesota, 22 October 1992.

10. *Management Practices: U.S. Companies Improve Performance Through Quality Efforts* (Washington, D.C.: Government Printing Office, Document No. GAO-NSIAD-91-190, 1991).

11. Kano, interview.

12. "World Business Rankings," *Wall Street Journal*, 24 September 1992.

13. George Lieb, "Strategic Planning and Business" (discussion group at the Quest for Excellence Conference, Washington, D.C., 7 February 1995).

14. Kano, interview.

15. *1995 Information Please Business Almanac and Sourcebook* (Boston: Houghton Mifflin, 1994), 289.

16. Ritz-Carlton (speech given at ASQC Section Conference, Pasadena, Calif., 19 January 1993).

17. Zytec Corporation, Showcase Handouts (presented at Zytec Showcase 92, Redwood Falls, Minnesota, 22 October 1992).

18. Executive Mission to Japan, October–November 1992. Includes data from Nippon Denso, Komatsu, Spa Resort Hawaiian, Nissan, Juki, Fuji Xerox, Y Hewlett-Packard.

19. Karen Bemowski, "To Go Boldly Where So Many Have Gone Before," *Quality Progress* (February 1995): 29.

20. Joe Nacchio, president of AT&T Universal Consumer Communications Services, interview by author, Washington, D.C., 6 February 1995.

21. John Huey, "The Leadership Industry," *Fortune*, 21 February 1994, 54.

22. John Huey, "The New Post Heroic Leadership," *Fortune*, 21 February 1994, 42.

23. Kano, interview.

24. Ronald Heinkoff, "CEOs Still Don't Walk the Talk," *Fortune*, 18 April 1994, 16.

Part 2

What to Do

Introduction

The steps of a total quality service (TQS) program are outlined to provide an easy-to-follow sequence of action. Any plan or process listed in bold has a companion tip sheet in Part 3: How to Do It. If you eventually want to apply for the Baldrige Award, the how-to section includes tips on those leadership behaviors, plans, measurements, systems, and processes stipulated in the Baldrige Award criteria.

Before the outline, you will find a TQS road map. This will give you a quick snapshot of the entire process. Each box on the map is detailed in Part 3.

Total Quality Service Road Map

Exhibit 2.1 illustrates the steps necessary to implement TQS. Finding the **champions** and forming a **steering committee** to guide the process is one of the first steps. Doing assessment through measuring customer needs is the second step. Combining those needs with competitor data in a **state-of-the-organization report** gives you a launching pad. This **state-of-the-organization report** provides a snapshot of your organization's finances, culture, personnel, products/services, and trends.

Almost immediately you need to get your training program in motion and set up a couple of starter improvement projects that are not too complex. Pick some quick fixes so you can showcase success-

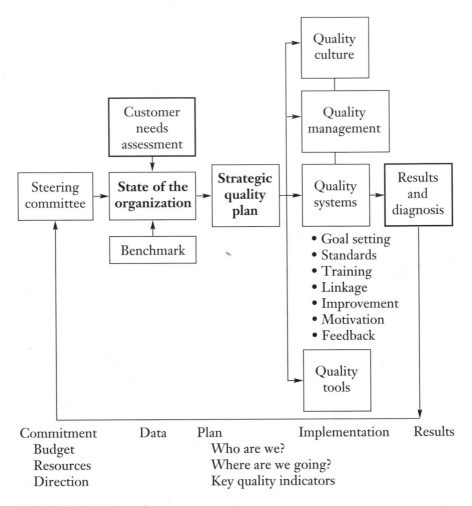

Exhibit 2.1 TQS road map.

es. These projects need to be linked to the organization's mission and vision. Early results will go a long way to convincing others that TQS can work within your organization. All this usually takes about a year.

After your **state-of-the-organization report** honestly tells you who you are, then you decide who you want to be. That is contained in the **strategic plan;** it spells out your goals—in target customers, products, and **key quality initiatives** to meet those goals. Many **strategic plans** are done every three to five years with specific objectives formulated every year.

The rest of this road map is implementation: simultaneously working on the culture, management, tools, and systems needed to sustain the TQS effort. Results are compiled at regular intervals on your **key quality initiatives** and are then given back to the **steering committee** for revised yearly goals and a report card on your overall **strategic plan.** Most large companies already do a **strategic plan.** Unfortunately, many don't know the difference between a business plan and a **strategic plan.** Many business plans provide some organizational and division goals with their appropriate budgets. A **strategic plan** identifies how you are going to strategically outfox the competition with your product or service. Companies have to compile data (done in a **strategic analysis**), be creative, and think about the future. To further complicate matters, several companies do a separate business, strategic and quality plan. George Lieb, vice president of strategic planning and business at GTE Directories, said that GTE used to do its business plan separate from its quality plan. At the 1995 Quest for Excellence Conference in Washington, D.C., Lieb talked about the integration being a powerful contributor to customer satisfaction and profits.

What-to-Do Outline

The outline summarizes the TQS process steps and provides a glimpse of the order of implementation. Each step is defined and explained in Part 3. Again, these steps have been consolidated as a result of working with or studying many of the Baldrige Award and Deming Prize–winning companies. Among them are Zytec, Motorola, Ritz Carlton, AT&T Universal Card Services, Nippon Denso, Eastman Chemical, Ames Rubber, AT&T Consumer Communications Services, Wainwright, Solectron, Juki, Y Hewlett-Packard, Fuji Xerox, and Spa Resort Hawaiian. My experience has shown that the right commitment and spirit behind service quality is more important than whether people follow the steps in order.

Many of the steps have titles that may be new to you. Each organization may have its own proprietary name for the plan, process, or tool involved. You might just glance through the chronological outline, then read Part 3, and come back and reread the outline. The definitions may be clearer at that point.

1.0 Leadership

1.1 Find the **champions** who will support the continuous improvement effort.

1.1.1   Identify where and who to find.
1.1.2   Determine what **champions** need to do during training and in pilot projects.
1.1.3   Ask, What if the **champions** are at the bottom of the organization?

1.2   Form a **steering committee.**
    1.2.1   Ensure success by attending to
        1.2.1.1   Criteria for selection
        1.2.1.2   Logistics
        1.2.1.3   Team building
    1.2.2   Do a **project plan for quality.**
    1.2.3   Do a **declaration of commitment.**
    1.2.4   Set up a **success story system.**
    1.2.5   Set up a **celebration system.**
    1.2.6   Develop a **visibility plan for results.**

1.3   Develop a **quality accountability system** including the following:
    1.3.1   Quality performance reviews
    1.3.2   Quality hiring tools
    1.3.3   Quality promotion tips

1.4   Develop guidelines and process for the **annual quality report.**

2.0   Information and Analysis

2.1   Select which key processes in business you want measured. See **strategic planning** and **key processes.**
    2.1.1   Select measurement devices to fit each information and analysis stage.
    2.1.2   Ask and answer the question, Once we set strategic and qulaity goals, how are we going to measure them?

2.2   Match which measurement devices are best suited to measure these **key processes.**
    2.2.1   Ask and answer the question, Once we get the measurements, how are we going to know how good that number is?
    2.2.2   Identify weaknesses.

2.3 Determine how you will use the data.

  2.3.1 Stipulate how you will use this information to improve the organization.

  2.3.2 Stipulate owners for each of the changes needed.

2.4 **Benchmark** other companies to get comparative data.

3.0 Customer Focus and Satisfaction

3.1 Determine **customer focus and satisfaction measurements.**

  3.1.1 Develop objectives for the **customer satisfaction measurement system.**

  3.1.2 Match tools to measurement objectives.

  3.1.3 Do **focus groups** or **interviews** to determine right questions.

  3.1.4 Do **customer satisfaction surveys** to determine quantitative and/or qualitative data.

  3.1.5 Develop other customer focus and satisfaction measures, feedback measures, observation, and/or **complaint tracking.**

  3.1.6 Develop ways of comparing your customer focus and satisfaction to your competition.

3.2 Have internal groups do an **internal customer focus and satisfaction** baseline.

  3.2.1 Have internal groups do systematic **focus groups** or **interviews** with their customers to find **key quality indicators.**

  3.2.2 Use these **key quality indicators** as questions on your **customer satisfaction surveys.**

3.3 Develop a **customer focus and satisfaction** implementation system.

  3.3.1 Measure (or baseline) current **key quality processes.**

  3.3.2 Develop intervals of measurement.

  3.3.3 Decide who is going to measure what and when.

  3.3.4 Decide how results will be communicated within the organization.

4.0 Strategic Planning

4.1 Do a **strategic analysis.**

4.1.1   Do a **state-of-the-organization report.**
4.1.2   Determine **industry trend analysis.**
4.1.3   Do **competitor profiles.**
4.1.4   Do a **strategic synthesis** (key selling points).

4.2   Do a **strategic plan.**
4.2.1   Write a **strategic plan** (target clients, goals, etc.).
4.2.2   Implement the **strategic plan.**
4.2.3   Review the plan at systematic intervals (monthly, quarterly, etc.).
4.2.4   Improve the strategic planning process at regular intervals.
4.2.5   Disseminate the **strategic plan.**

4.3.   Do a quality analysis.
4.3.1   Determine **key processes** in your business.
4.3.1.1   Discern how results are currently measured.
4.3.1.2   Decide where you will measure **key processes** (which departments, suppliers, etc.).
4.3.1.3   Decide on appropriate number and intervals of measurement.
4.3.1.4   Flowchart each of the **key processes** using a team.
4.3.1.5   Identify areas for improvement and begin improvement teams.
4.3.1.6   Baseline those **key processes**—measure *before* improvements are implemented.
4.3.2   **Benchmark** key processes in organizations that have the "best in class."
4.3.3   Determine **key quality objectives.**

4.4   Develop a **quality plan.**
4.4.1   Write a **quality plan.**
4.4.2   Implement the **quality plan.**
4.4.3   Review the plan at preset intervals, such as daily, weekly, monthly, or quarterly.
4.4.4   Improve the quality planning process at regular intervals.

4.5   Develop a **procedures manual.**
4.5.1   Have various departments write job procedures.

4.5.2 Do a relevant ISO 9000 internal audit.

5.0 Human Resource Development and Management

5.1 Develop a **human resource and management plan,** which becomes part of the **strategic plan.**

  5.1.1 The **human resource and management plan** includes the following:

   5.1.1.1 High-performance work systems

   5.1.1.2 Employee development

   5.1.1.3 Recruitment

   5.1.1.4 Mobility, flexibility, and changes in work schedule

   5.1.1.5 Reward, recognition, and benefits

5.2 Develop an **employee involvement process.**

  5.2.1 Create a **suggestion system.**

  5.2.2 Initiate **quality improvement teams.**

5.3 Refine the **employee education and training plan.**

  5.3.1 Include human resource planning.

  5.3.2 Review employee involvement, **suggestion systems,** and **quality improvement teams.**

  5.3.3 Develop the **employee education and training plan** including the following:

   5.3.3.1 Needs assessment

   5.3.3.2 Linkage to short-term and long-term plans

   5.3.3.3 On-the-job training

   5.3.3.4 Classroom training

   5.3.3.5 Measurement of results

  5.3.4 Determine the **employee performance and recognition** system.

5.4 Investigate **employee well-being and satisfaction measures** by looking at the following:

  5.4.1 Employee satisfaction and improvement goals and measures

  5.4.2 Quantitative measures (safety, turnover, equal employment opportunities, absenteeism, and accidents numbers)

5.4.3 An employee satisfaction improvement plan

5.4.4 Perceptual measures (surveys, focus groups, and advisory groups)

6.0 Process Management

6.1 Develop design and modification of products or services system. Tools like **quality function deployment,** where client needs can be matched to specifications, can be very helpful.

6.2 Refine processes (production and delivery) and quality management system. This includes calibration of instruments.

6.3 Refine the **support services quality system.**

6.4 Refine the **supplier quality management system.**

6.5 Refine the **quality assessment system.**
5.5.1 Determine how assessment systems are used to improve services/products.
6.5.2 Determine how the company verifies that assessment leads to actions.

7.0 Business Results (Look for level, trend, and variability.)

7.1 Report product/service quality results.

7.2 Report company operational results.

7.3 Report business process and support service results.

7.4 Report supplier quality results.

Summary

This what-to-do part outlined the chronological steps common in establishing a TQS program. Each element will be explained in the next part. The steps actually happen in a chaotic, holistic fashion. Inspiring employees to embrace each step requires both flexibility and creativity. As noted, it might be helpful to reread this chapter after you have gone through a full explanation of the steps.

Part 3

How to Do It

Introduction

Part 2 provided simplified and unexplained chronological steps. This part explains implementation and provides examples, a sample table of contents for some of the reports, and caveats about what doesn't work well. The illustrations and tips come from Baldrige Award–winning, Deming Prize–winning, or other high-performance companies I have worked with or benchmarked. This section is organized the same way as the Baldrige Award criteria is arranged to make eventual application easy.

This how-to part is meant to be modified to fit your business or group. The material is designed to give you a starting place. You can pick and choose which tools, processes, reports, and plans you want to use as your launching pad.

Since Baldrige Award–level performance is the ultimate goal, a first glance of all that is required may be intimidating. This is reduced when you realize that everyone in the organization eventually gets involved. You train people in the basics and then delegate the implementation. Thus, one group will not have the responsibility for all of the steps. Each part of the organization will own its separate piece. The function of a **steering committee** is to have a group to coordinate all of the pieces.

Implementation of all that follows may take several years. AT&T Universal Card Services took 18 months because it designed the organization from scratch. From my own experience, retrofits take anywhere from two to five years. Timing will depend on your **declaration of commitment** from the top and how well you have recruited customer-focused people in the past.

1.0

Leadership

Introduction

During the 30 years he worked in Japan, Dr. W. Edwards Deming shifted his focus from statistical process control to leadership. According to the late Deming, quality begins and ends with leadership. Without this vital ingredient, quality is confined to little "bits and pieces" throughout the organization. According to Kano, Japan's leading expert on quality, bits-and-pieces quality was exactly where Japan started in the 1960s.[1] Leadership was the vital ingredient that led Japan to become an economic powerhouse in the late 1980s.

The Baldrige Award examiners look for leaders who are actively involved in promoting quality, in establishing the **strategic plan**, in determining which tools provide measurements for **key quality indicators,** and in figuring out how to increase both employee and **customer focus and satisfaction.** "Blowing smoke" doesn't work with Baldrige Award examiners. They look for the intent to provide customers with extraordinary products and service. The examiners also make sure that the whole cast of characters at the top is involved—not just the CEO.

What if you don't have a commitment to implement quality at the top? You're perfectly normal. Quality consensus usually starts with a few people. The key is to get a critical mass as far up the organization as possible. Your naysayers will either climb on board or leave. My discussions with the Japanese companies revealed that they, too, went through this evolution and have no magical answers. Gradually, critical mass evolved through compliance, attrition, or a swift kick out. For the Japanese the kick was usually to some quiet spot without responsibility. At least they didn't kick their obsolete managers upstairs.

Middle management buy-in is usually the most difficult leadership hurdle. Most quality organizations have pancaked themselves into just a couple of layers. The problem goes away when this happens. Middle managers have the most to gain by resisting, since they are the ones that have to give up control in the empowerment process.

Companies that empower employees as part of their overall total quality management effort are twice as likely as other firms to report significant product or service improvement.[2]

Several studies have cited a lack of U.S. management readiness to undergo the rigors of rapid change. One Gallup Poll revealed that more than 75 percent of the 400 executive respondents said that there were four main reasons why executives resist change.[3]

1. They have a vested interest in the status quo.
2. They don't know what to do about change.
3. They don't like to lose control.
4. Their vision and goals are too short term.

Thus, a **TQS culture** conducive to continuous improvement is a starting point. Continuous improvement, or *kaizen*, is approached with religious fervor in Japan. Kaizen looks for positive ways to acknowledge progress, not negative ways to punish people for not meeting goals.

The leaders are where kaizen starts. They model self-improvement. This attitude is vastly different from the traditional, "We're here because we know it all. That is why we get paid the big bucks." TQS leaders don't know all the right answers; they know all the right questions. Quality-oriented leaders go to conferences and seminars with a hunger to learn what is the best management practice. Note that Japanese CEOs go to seminars. JUSE has been a focal point for the top Japanese brass to learn from each other. Deming talked about having 80 percent of Japanese wealth represented in some of his seminars.[4] He had a seminar expressly for CEOs in the United States in 1992, where the 120 who attended were mainly high-level managers, not CEOs.[5] Highlight this paragraph and drop the book on your CEO's desk if this is a problem at your organization. Top executives must understand that the U.S. economy is doomed if they don't switch from a know-it-all culture to one where it is okay for them to learn from empowered employees at the lowest ranks and from the outside. There are powerful examples of new leadership, necessary in the 1990s, at Federal Express, Marriott, Motorola, Milliken, and AT&T Universal Card Services.

Remember, leadership also needs to permeate throughout the ranks. It does so through carefully selected quality **champions** and a well-functioning quality **steering committee**. Tips on selection and execution of both are detailed in the following sections.

Total Quality Service Culture

Paradigm Shifts

Exhibit 3.1 shows paradigm shifts that underlie a **TQS culture.** A paradigm shift is a fundamental change in the way one views the world. The book and video *Paradigm Shifts* elaborate on this theme.

Champions

Quality is impossible without **champions** in the organization. The higher those **champions** are in the organization the better. Find open, curious, smart, and articulate people and send them to a quality training seminar. Then identify them as your quality **champions**.

The Japanese call their **champions** "quality promoters." Quality promoters typically have 25 percent of their job description dedicated to promoting quality within the ranks.[6] These are not bolt-on quality staff jobs, but are additional responsibilities given to those who want to be promoted within the organization. Being a quality promoter is a highly valued honor.

The **champions** are frequently trained as quality facilitators and are part of the cascade training that Xerox described in its implementation process. Outsiders train the facilitators, and the facilitators train others in the organization. Xerox, a 1989 winner of the Baldrige Award, used this cascade process to train over 100,000 employees in four years.[7] **Champions** were carefully chosen to carry the message.

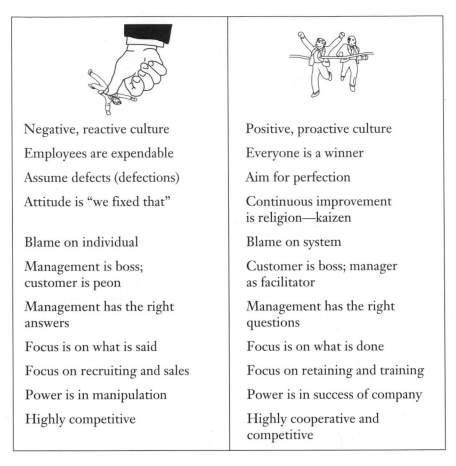

Negative, reactive culture	Positive, proactive culture
Employees are expendable	Everyone is a winner
Assume defects (defections)	Aim for perfection
Attitude is "we fixed that"	Continuous improvement is religion—kaizen
Blame on individual	Blame on system
Management is boss; customer is peon	Customer is boss; manager as facilitator
Management has the right answers	Management has the right questions
Focus is on what is said	Focus is on what is done
Focus on recruiting and sales	Focus on retaining and training
Power is in manipulation	Power is in success of company
Highly competitive	Highly cooperative and competitive

Exhibit 3.1 Paradigm shifts.

Steering Committee

A quality effort in any organization with more than 15 people needs a **steering committee,** since any group with more than eight members usually has difficulty with decisions. You want a **steering committee** to guide your effort in the organization. The **steering committee** also sets the pace for implementation and accelerates buy-in throughout the organization.

Typical actions of the **steering committee** include the following:

- Appoints quality **champions.**
- Develops the **project plan for quality.**

- Determines the **employee education and training plan.**
- Determines **suggestion system** and **success story system.**
- Recruits and trains facilitators and trainers within the organization.
- Takes **key quality objectives** and recruits **quality improvement teams** to tackle solutions to those questions.
- Develops a **celebration plan.**

Selection

Selection for the **steering committee** is frequently based on a diagonal slice of the organization. The benefits of a diagonal slice rather than just a top tier are as follows:

- The **steering committee** can include powerful influencers from low levels of the organization.
- If just high-level people are chosen, the group looses its ability to get a frontline perspective.
- The few low-level positions can be used to give outstanding promoters visibility in the organization and can reward them for quality-related attributes.

Various levels and departments need to be represented. Some companies have both an executive and a quality steering committee. The executive group is a horizontal slice off the top and the quality group is more diverse. Many companies that started by selecting titles to go on the **steering committee** found that they were better off being more selective about the personalities they chose. In addition to the person's position, the following criteria might also be considered.

- Personality—open-minded, energetic, and articulate
- Cooperative—willing and able to work in a team
- Influence—able to influence others in the organization

Rather than make decision making unwieldy by having too many people on the **steering committee,** call in the experts as they are needed to give advice on particular issues. A reasonable committee size will keep the decision process manageable.

Execution

During the first few months of a quality effort, the **steering committee** might want to meet once a week. Subsequent meetings can be at longer intervals as the process matures. Intervals need to be both frequent enough to get the process moving and not so frequent that members complain.

Team-Building Exercises

Team building within the **steering committee** is a critical starting point. Team-building exercises can be of particular benefit. Having an outside facilitator during the early stages can help the team learn good problem-solving skills and move forward quickly.

Project Plan for Quality

Many organizations flounder for months or sometimes years when they first start a quality effort. A **project plan for quality** can set realistic expectations, communicate intentions clearly, and allow you to easily monitor progress. Exhibit 3.2 provides a template for a **project plan for quality.** It allows the committee to detail who will be responsible for specific elements of implementation. Each of the elements mentioned in the systems area (visibility plan, celebration plan, and so on) are detailed in sections under those names.

Declaration of Commitment

A **declaration of commitment (DOC)** is merely the taking of a vote to determine what level of commitment the **steering committee** wants for the quality effort. The benefits for evaluating commitment mean that the **steering committee** can

- Clarify the urgency or pace of deployment.
- Allocate resources (time, people, and money) accordingly.
- Send a clear message to the organization on the sense of urgency.

Steering committee members	Mission of steering committee
1.	
2.	
3.	
4.	
5.	
6.	
7.	
8.	

Goals of steering committee		
What	**Who**	**When**
1.		
2.		
3.		

Systems plans*		
What	**Who**	**When**
Visibility plan for results		
Training plan		
Suggestion system for employees		
Celebration system		
Customer needs assessment system		
Employee satisfaction assessment system		
Quality improvement teams		
Other _____		

*Systems need to include who, what, when, and how, plus a flowchart of how they will unfold. A schedule is also useful. Remember to make those systems visible to all by using bulletin boards, newsletters, and upper-level managers' announcements.

Exhibit 3.2 Project plan for quality steering committee.

- Develop realistic expectations about the length of the transformation. Honesty is imperative if the effort is credible.

Take a secret ballot. Have each individual vote on a number that corresponds to his or her level of commitment. The choices are listed. Beside each choice is what you hear when you eavesdrop on the true commitment being expressed in the hall:

1. Forget it—"We're fine the way we are."
2. Awareness—"We will focus on employee awareness of quality."
3. Slow growth—"We will start training and teams, and will worry about results later."
4. Go for the silver—"We will train everyone within 18 months, and set stretch quality goals."
5. Go for the gold—"We will do a rigorous self-assessment and apply for the Baldrige Award within one or two years."

Caveat: A secret ballot is important when the culture of the organization has emphasized authority and compliance. If the vote is by hand, the compliant group will all look to the leader and vote accordingly.

How to Use the DOC

The DOC tool is primarily used for internal clarification. If the group chooses to go for the silver or gold, make sure enough resources have been allocated. Extensive training and time will be required to achieve either of these goals. Small companies can get by with less cost because informal communications systems can sometimes substitute for formal and costly mechanisms to distribute knowledge.[8]

Tell employees of the **steering committee's** intentions. If you're going to take a cautious approach, say so. Going for the gold will invite resistance among the managers who are threatened by the change. Understand that you will have to deal with that resistance.

At an executive retreat for one of my clients, an engineering company, a couple of managers said that they didn't have time for all the quality activities because they were overworked in their regular jobs. At a break, one of them put the following on the board: "Quality is in the way."

The CEO returned from the break and promptly erased one word so the missive read; "Quality is the way." Signaling clear intentions to managers helps them climb on board or get out of the way.

Success Story System

Success stories must be gathered immediately to help fan the fires of quality. These success stories can be ones that happened before the TQS movement began. They illustrate that continuous improvement is already part of the system. A **success story system** sets up a centralized repository for success stories and has a plan for communicating those successes to employees, customers, suppliers, and even the outside world. The coordinator of the system helps set direction and monitors and disseminates the stories.

The benefits of a **success story system** include the following:

- Builds morale.
- Motivates employees to go the extra mile.
- Motivates clients to buy your products and services.
- Educates employees about what other departments are accomplishing.

How to Implement the System

A single person in the organization needs to be identified to do the following:

- Solicit success stories
- Provide a template for reporting them
- Disseminate the stories through the following:
  —Showcases or quality forums
  —Conferences
  —Bulletin boards
  —Newspaper articles
  —Recognition celebrations
  —Incessantly talking about them
  —Congratulations from the CEO, upper-level managers, and/or management by walking around

David Kearns, CEO of Xerox when it won the Baldrige Award in 1989, talked about how important the Teamwork Days were in showcasing improvements. The quality fairs were attended by employees, customers, suppliers, and community and academic leaders. Over the years the fairs grew in popularity. They served to motivate and inform employees and allowed management to "communicate with employees in an upbeat way."[9]

Departmental quality **champions** are responsible for getting success stories from their own groups. The person who coordinates success stories, though, needs to assume a central pivot point so marketing knows where to go. This should be a part-time function of someone like the vice president of quality or the training director.

A Visibility Plan for Results

Unless most organizations focus on making the intangible tangible, it doesn't happen. One of the striking differences between companies that have reaped the benefits of continuous improvements and those that have not is the visibility of their efforts. Americans' problem-solving heritage has made it difficult for many employees to take credit for internal progress. Giving visible credit is a fundamental cultural change that needs to happen for sustained success.

A **visibility plan for results** is a conscious effort to make results recognizable. It starts with collecting success stories. Quality **champions** are the ambassadors who make the successes visible. They become the troubadours to informally share relevant successes with coworkers. **Champions** may have formal responsibilities to make sure the bulletin boards are refreshed with success stories.

A study done by Brooks Carder and James Clark at IBM revealed the importance of recognition programs. Seventy-seven percent of the respondents of a survey said that recognition efforts improved the performance of employees. Yet 59 percent of the administrators felt that the number of people recognized was too low. The administrators said that it was difficult to go to meetings where only sales and marketing people were recognized.[10] This is consistent with my own experience with engineering, broadcasting, insurance, computer, software, and manufacturing companies.

Regular meetings with quality **champions** help build involvement and motivation. The **visibility plan for results** helps the **steering**

committee concentrate on who is going to write the news articles, who is going to change the bulletin boards, who is going to feed the CEO information on worthwhile successes, and who is going to organize showcases. Each of these avenues for visibility needs to have standards (change bulletin boards every 10 days, for instance) and sometimes a budget. One client of mine even had a neon sign that helped announce recent successes. AT&T Universal Card Services follows its performance on 110 measures that are updated daily and posted for all employees to see.[11]

During my 1992 executive mission to Japan, I was intrigued by the sophistication of the marketing of quality in the Deming Prize–winning companies. Examples of the companies included Nippon Denso, Nissan, Toyo Engineering, Spa Resort Hawaiian, Juki, and Komatsu. These organizations were not offensive or even obvious about the way they sold themselves. They all used the same powerful approach. Results were highly visible on the bulletin boards in the factories and offices. Representatives throughout the organization were conscripted to talk about their group's evolution and successes. Just the mere act of engaging employees in selling themselves to each other and the outside created pride and energy to keep on improving. They all followed much the same outline in describing their quality journeys. It included the following:

- A brief history of the company including original products and target markets.

- The motivation for getting involved with quality. For many it was being on the brink of bankruptcy or an immediate competitive threat. In the 1950s Japanese motivation included a worldwide perception that "Made in Japan" was a quality joke.

- The first steps, difficulties of start-up, and progress thereof.

- Graphs of profits, revenues, and numbers of employees before and after implementation.

- Descriptions of how the company achieved its goals; for example, through **leadership, champions, strategic planning,** deployment, training, tools, employee involvement, motivational campaigns, monitoring, and rewarding efforts.

- Quality objectives for the next few years.

Descriptions were honest and included the difficulties as well as the successes. Over a period of 40 years these Japanese winners have been through much trial and error and were willing to share both. The power of this substantive sales approach was to leave visitors with the understanding that the quality imperative was as revolutionary a change as the invention of the electric light bulb. Quality has changed the global landscape as much as electricity has.

Deming had extreme difficulty listening to U.S. managers say, "It works for the Japanese, but not for us" or "It's just another fad, it will go away."[12] He knew that these statements meant that U.S. economic dominance would go away, but not the quality movement. Thank goodness many U.S. service company executives finally understood that the world was changing in a profound way before their eyes.

Use Exhibit 3.3 to organize your visibility plan.

What	Who	Timing	Resources
Bulletin boards			
News articles			
Executive pipeline			
Showcases			
Neon signs			
Other			
Other			

Exhibit 3.3 Visibility plan for results.

Celebration System Tips

A **celebration system** consists of both planned and spontaneous ways to celebrate individual and team successes. The **steering committee** usually comes up with the large events and each department is empowered to come up with its own small ways of recognizing success.

Brainstorm ways you can creatively recognize team and individual successes. You will need to consider the following:

- Criteria: What determines who will receive special recognition?

- Award: Will the award be money, goods, or an experience?

- How dramatized: How will the celebration unfold? Will there be a large audience? Should it be included in the service company or local newspaper?

- Who will award: Who will hand out the certificate or award? The higher the level of the person giving the award, the more prestigious the award. Secretaries, maintenance people, and truck drivers are just as deserving of CEO attention as are salespeople.

- Balance: What balance will exist between team and individual awards?

- Who will plan: Who will do the work of planning the award ceremony?

Tips on Celebration Systems

1. The criteria should be clear to all. Award winners need to be praised for the particular aspect of the criteria they attained.

2. Experience is one of the most powerful motivations. Incentive trips are not welcomed by everyone. Money is nice but becomes its own tyrant if a system is based on money. The Japanese use an annual quality conference as a key motivation. Individuals and groups who excel are invited to this conference to showcase their results. The perk helps educate the rest of the company and country on new methods. It also gives people visibility and has an incentive trip built in. Organizations from Xerox to Fox Valley College in Appleton, Wisconsin have quality fairs to celebrate team successes.

3. Celebrations are best planned internally. Outside companies can help facilitate, but creativity within the TQS company is a powerful resource.

4. Ask employees how they want to be rewarded. The best ideas come from inside.

5. Regular intervals of recognition are usually not as effective as recognition based on real successes. Too many employee-of-the-month systems lose employee respect because the manager is forced to invent success because it is the first of the month.

6. Peer-determined successes are better than just manager-determined successes. Managers don't know who has really done the work. Let employees recommend each other for acknowledgment.

7. Celebrations seem to need free food. It doesn't have to be much. Even free ice-cream cones are okay. Elaborate and fancy food can be expensive and is unnecessary.

8. Certificates of appreciation are usually as good as plaques. Few people put plaques on their walls so they become a storage question. Ask about your employees' preferences in whether they prefer a plaque, paperweight, certificate, or other symbol of achievement.

9. The way celebrations are done is just as important as the fact they happen. Put extra creativity and fun into the presentation and announcement of the award.

10. Informal recognition is even more sustaining than banquets. Many companies have a system such as "Catch Someone in the Act of Doing Something Right." Zytec has employees give each other colored beads when a coworker goes above and beyond the call of duty.[13] Rancho Vista Bank in Vista, California uses a yellow rose symbol. Employees will lay a yellow rose on a coworker's desk when they particularly appreciate another employee's deed.[14]

Starter List of Celebration Ideas

1. Quality fairs where various groups and individuals showcase their success.
2. Annual quality awards banquet.
3. A day off. Successful individuals get to choose the managers to replace them on their day off.
4. Trip to share the ideas or methods with another division.
5. Award luncheons.
6. A presentation at the next board of directors meeting.
7. One department hosts a celebration party for another successful department.

8. Trips to benchmark other companies.

9. Notes from the president on a job well done. Employees used to get "buck-o-grams," a congratulatory note, from Buck Michael when he was CEO of Fluor Daniel.

Quality Accountability System

Many companies spend thousands and sometimes millions of dollars training their employees in quality only to find out they don't follow the precepts when back at their jobs. The problem is accountability. Superficial programs get superficial results. Employees need to have training linked to achieving measurable outcomes and the vision of the organization. Accountability needs to be threaded through customer needs, company vision, training, improvement efforts, and measurement. Many companies thought that just training would transform their organizations.

The successful U.S. companies with which I have worked and the Japanese Deming Prize–winning companies I have benchmarked emphasize *positive accountability*. This focuses on goals that have been met, improvements that have been made, and individuals who have put forth special efforts. The less successful companies either have little or negative accountability. They focus on groups who haven't met their goals and individuals who are falling down on the job. Individuals don't make many suggestions because they are afraid of getting singled out as malcontents. The Japanese companies revised their systems to support positive accountability. Those systems seek improvement-oriented and successful people in hiring, promotions, compensation, and performance reviews. The following section provides some tips on how each of these systems can support your quality effort.

Hiring Tips

Reference Checks
Assess an applicant's quality orientation in a neutral manner. Listen for specific results, the ability to work on a team, and the ability to give credit to others. Use statements such as the following:

- Describe Person X's preference in working independently or on a team.
- Talk about how Person X gives recognition and credit to others.

- What were Person X's accomplishments in quality?
- What did Person X do to increase customer satisfaction?

Interview Questions

Ask the potential employee questions that relate to quality. Examples include the following:

- What are your continuous improvement success stories?
- What individuals or companies do you most respect in continuous improvement?
- What books or literature have you read in quality?
- What aspects of quality do you find important on your job?

Look for specifics in the applicant's answers, the ability to work on a team, a customer orientation, and job loyalty. Beware of people who have changed jobs every two years. You don't have much accountability in short-duration jobs. Watch also for someone who does an excellent job of promoting him- or herself and not others.

Well-done reference checks or simulations can be more reliable than interviews. Other companies have used job simulations to help assess. AT&T Universal Card Services uses a telephone interview to determine if its applicants have a smile in their voice.[15]

Performance Appraisal Tips

1. Each person in the organization should have a set of goals that relate to the overall strategic quality objectives. Make sure these are part of the review process.
2. Use input from both peers and clients of the employee. AT&T Consumer Communications Services uses a 360° review all the way up to the president's level.[16]
3. Have employees write their own performance appraisals. You will learn a lot about their accomplishments and goals by using theirs as a start.
4. Don't rank order employees. Deming was strongly opposed to this common practice. His point was clear. Why make 90 percent of your employees feel like losers? Your recruitment system should be strong enough to assume that you have 100 percent winners.

Compensation Tips

1. Commission-only sales systems do not promote customer service. Notice how Saturn, Lexus, and Infiniti have redefined the quality of sales by putting their sales force on salary. James McIngvale, who owns the Gallery Furniture Store in Dallas, attributes a 40 percent increase in profits in 1991 and 1992 to a switch from commissions to salaries.[17] A radio broadcasting TQS company client has switched its tradition of commission-only sales pay to predominately salary plus bonus. The company found that salespeople were negatively motivated for service on commission only. The revenues went up more than 150 percent after the switch. The organization also found, however, that switching from commission to salary plus bonus involved self-motivated and sophisticated salespeople, training, and attention to the transition. The benefit of switching was that it was easy to implement customer focus and team selling.

2. If you have a bonus system, apply it to all employees. Wainwright, winner of the 1994 Baldrige Award, takes the yearly profit and splits it equally among all employees.[18] Many companies make the mistake of concentrating bonuses on just their sales force or upper-level managers. You lose leverage in your system if you have imbalances. Saturn rewards employees on quality, productivity, and profit.[19]

3. Other companies pay employees based on skills. Milliken pays its associates on the basis of skills they have acquired. Seniority depends on skills, not years.[20]

Annual Quality Report

Most every company has an annual report. Usually it focuses on new strategies and gives the financial results of the organization. Total quality management or service puts the focus on quality and assumes that the finances will follow increased sales. Many Japanese companies also have an **annual quality report** that receives just as much attention as does the financial one. North Federal Credit Union of San Diego, winner of a Silver Eureka Award from the California Society of Quality and Service, also produces a short **annual quality report** that it circulates among employees and customers.[21]

The benefits of an **annual quality report** include the following:

- Helps upper-level managers focus on improvement.
- Helps make quality data visible.
- Provides a valuable marketing tool if you have positive results.

How to Do an Annual Quality Report

1. The **strategic plan** should contain the **key quality initiatives** for the near term (one year) and long term (three years or beyond).

2. The **annual quality report** contains results from those efforts and acknowledges the special efforts by different divisions, departments, or people.

3. The cost of savings can be reflected in this report as well. Cost of savings might include such measures as savings from reducing scrap, from increased productivity, from increased business due to customer satisfaction, and/or from reducing cycle time.

4. The review process for the **annual quality report** is critical when driving a quality effort. In Japan, CEOs or division heads spend 30 percent to 50 percent of their time going to presentations, reviewing results, and removing obstacles in the way of achieving those results.[22]

Caveats:

1. When companies first start the **annual quality report** and review process, their managers typically use a well-rehearsed critical management style. Goals are not always met. Managers need special coaching on how to provide constructive feedback and how to change the focus from the individual to the contributing system's problems. The focus needs to be on how the next level up can change a system, remove an obstacle, or provide a resource so the goal can be met.

2. Quality goals must be negotiated. I heard this at every Japanese company I visited. The Japanese started this process by delegating quality goals. They learned quickly that buy-in was poor and excuses were abundant with top-down goal setting.

2.0

Information and Analysis Tools

Introduction

The Baldrige Award allocates the Information and Analysis section only 75 points out of 1000. Yet, this section drives the entire process. Baldrige Award examiners look to see that what is measured is actually driven by customer needs and the **strategic plan.** Is the organization focused on its **key quality objectives** or just collecting meaningless and unused data?

Total quality management (TQM) requires data and analysis. JUSE counselor Kano described the acronym that the Japanese use for "gut-level management." It is called KKD management, where

Keiken = experience
Kan = intuition
Dokyo = guts

Thus, one of the huge before-and-after differences in TQM is the use of data rather than just KKD. Baldrige Award examiners want to see that the data are used for planning and day-to-day management. They want to see that data are used to ensure timeliness, reliability, and other quality objectives.

Criteria for Data Collection

Data must be relevant, reliable, and representative. *Relevant* data mean you need to choose what you measure based on what is important to your customers and the quality of your service or product. GTE Directories is a good example. It learned what customers wanted *after* needlessly spending $15 million on a new press. Marilyn Carlson, vice president of sales at GTE, said that they didn't start their market share turnaround until customers designed their own questions.[23]

Reliable data mean that you can count on the results to be the same the next time you give the survey or measure the results. Unreliable data are worse than none at all. Questionnaires that have ambiguous questions or combine many questions into one will have inconsistent results. Questionnaires that are given just after people have been waiting in line a long time may skew the results. One engine manufacturing company I worked with found that its customer satisfaction surveys were more reliable if they were sent out a week after the completion of the service instead of when the customer picked up the engine.

Representative data require that you use statistical techniques to either randomly sample your customers, widgets, suppliers, or whatever you are measuring, or sample the total population. Just collecting data from a subset (like customers who complain) means that your data may not be representative. One major airline with which I worked didn't realize the importance of this element when it asked passengers about food preferences. The airline surveyed 15 business passengers from New York and found them to prefer pasta. When the airline started serving pasta in first class, the health-conscious California business passengers started complaining. Results from 15 New York passengers cannot be extrapolated to cover millions of passengers who are much more diverse.

Key Considerations in Data

Being data driven is not enough. Data first must be useful, meaningful, and important. Then, data must be visible to everyone to make a difference. Japanese companies have kaizen corners. Kaizen corners are areas of the factory or facility that showcase charts that reflect continuous improvement. These charts are updated frequently—usually no less than twice a month. At Nissan, Juki, Komatsu, and Toyota you see those graphs (see **control charts**) at each team's workstation. Employees see their own progress, and the charts can serve to help employees monitor and motivate themselves.

Putting data in user-friendly form is another key to success. Bar charts, **Pareto charts,** and **control charts** are means to make complicated data easy to interpret. Make sure that the data are both visible and in graphic form.

Baldrige Award examiners will want to see comparative data and how you put it to good use. Mark Graham Brown in *Baldrige Award–Winning Quality* reports that examiners look for three main things about the results that your data reveal.[24]

Level:	Is the overall level of satisfaction in the 90th percentile or the 50th? What is the level compared to benchmark data?
Trend:	Are the data going in a favorable direction? Is customer satisfaction improving and the number of defects declining? Are these trends sustained over at least three years?
Variability:	Do the data reveal a predictable pattern or are they wildly all over the chart?

Data alone do not tell you much. How do your data compare to that of the competition? How do they compare to your own goals? The Baldrige Award criteria also requires that you compare yourself to industry standards. Benchmarking, or studying best-in-class processes at other companies, plays a role in amassing comparative data.

Most importantly, are the data used to improve the company or to just sit on the shelf? What systems are used to incorporate the data into new goals and improvements?

Types of Information and Analysis Tools

Types of measurements vary. Basic categories and their tools include the following:

- Customer needs assessment and satisfaction tools
    - **Quality function deployment**
    - **Focus groups**
    - **Customer satisfaction surveys**
- Statistical process control (SPC) tools
    - **Control charts**
    - **Fishbone diagrams**
    - **Check sheets**
    - **Pareto charts**

> —Histograms
> —Scatter diagrams
> • Team process tools
> —**Flowcharts**
> —The quality story
> —Brainstorming

Samples of the most popular SPC tools are included here for quick reference. Customer focus and satisfaction tools are all covered in the appropriately titled section. Because several other books, like Nancy Tague's *Toolbox*, detail SPC and process tool formulas very well, this book only reviews them. *The Link* by Nevienne Torki is one of the best. It is a good reference book on the formulas that are used in some of the SPC tools.

Caveat: Many times the tools or data analysis devices become the master of the process. Make sure that the people are the masters and the tools are the slaves. One division manager at Komatsu in Japan talked about how its engineers used to brag about how complicated their **quality function deployment** (QFD) charts, which tied client specifications to features, had become. Groups would compare their large charts to tsunami mats and compete on how many tsunami mats were created by their QFD charts. They had to reign in the engineers and refocus on the purpose of using them to create happy customers and profits.

Control Charts

A **control chart** is a managerial tool that is used to monitor a process to see whether it is in statistical control. A certain amount of variation is normal in manufacturing and service processes. Sometimes phones get answered in one ring, sometimes in six. Sometimes a rod comes off a manufacturing line at 4.00002 feet and sometimes at 3.998 feet. The upper control limit (UCL) and lower control limit (LCL) indicate how much variation is typical for the process. Over 99.97 percent of the results (three standard deviations or three sigma) fall within the UCL and LCL. Points that fall outside those limits indicate special rather than common variation.

Exhibit 3.4 illustrates a **control chart** that plots hours missed. As you can see, hours missed varies between 250 and 750 hours a month for that work group. That amount of variation is common.

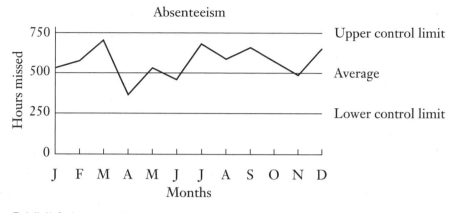

Exhibit 3.4 Control chart.

Control charts are some of the most popular tools used in both manufacturing and service firms. The concept behind the charts is critical to total quality manufacturing or service. **Control charts** help identify what the current practices and systems can predictably yield. A bank may find that errors on statements range from 120 to 200 defects per facility per month. Over time over 99 percent of the facilities reveal that error rate. A **control chart** would show that over 99 percent of the error rates fall between 120 and 200 defects. If managers examine each monthly error rate, they will find that the system can get worse. If managers don't like the range, they need to change the system. Hiring and training practices may need to change. The computer system may need to change. Management that focuses on that 120–200 defect rate and reacts to every slight variation further disrupts what is a stable system. Managers need to work the special causes of variation and change the systems so they yield precise and predictable results.

Fishbone Diagrams

A fishbone or cause-and-effect diagram is an easy way for groups to visualize complex reasons for a problem or possible solutions for it as they brainstorm. Exhibit 3.5 illustrates a simple **fishbone diagram.**

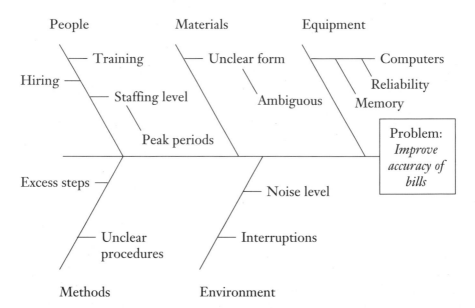

Exhibit 3.5   Cause-and-effect or fishbone diagram.

How to Construct and Use Fishbone Diagrams

Assemble the appropriate team. That means that you need the experts in the room and those involved with implementation. Use a flip chart or large piece of paper that everyone can see. State your problem or opportunity in the box at the end of the main fish bone. Develop your main categories, such as people, measurement, and so on. Then brainstorm subcategories for each.

The **fishbone diagram** is a popular tool to begin to look for root causes of problems. It allows groups to organize complex information and pick areas that are worthy of further investigation. Perhaps the data from the **customer satisfaction surveys** reveal that customers are not happy with the turnaround time from purchase to billing. Employees can use the **fishbone diagram** to brainstorm an opportunity; for example, how to improve billing cycle time. The free-floating thoughts can sometimes uncover causes that aren't apparent when superficial analyses are done.

After the discussion is over, go back and vote on the top five priorities in either your problem definition or possible solution. Then go out and test these priorities against real-world data using a **check sheet**.

Check Sheets

Check sheets are forms on which to record data. For example, an insurance company I worked with had trouble with billing errors. A group got together and brainstormed possible causes in a **fishbone diagram.** Four possible reasons for the errors were identified.

1. Inaccurate claim number
2. Inaccurate addresses
3. Incomplete forms
4. Illegible handwriting

The next step was to actually count the source of the errors across the various offices. A **check sheet** was used to list how many of each type of error existed. The company found that the incomplete financial information was the number-one source of errors. It turned out that salespeople had not understood some of the categories and had been using inconsistent information.

These self-designed tools need to be easy to read and interpret. Just use check marks or ticks to indicate any incidence. Exhibit 3.6 shows a **check sheet** that has not been completed; Exhibit 3.7 shows a completed **check sheet.**

Time period: _____ Department: _____ Purpose: _____ Name: _____ Date: _____		
Problem/Error	Frequency	Total

Exhibit 3.6 Check sheet.

Time period: January, 1993		Department: Accounting
Purpose: Causes of invoice error		Name: Ron Farber
Date: Feb. 2, 1993		

Problem	Frequency	Total
1. No invoice number	⊬⊬⊦ ⊬⊬⊦ ⊬⊬⊦ ⊬⊬⊦ ⊬⊬⊦ Ⅰ	26
2. Misspelled name	⊬⊬⊦ Ⅰ Ⅰ	7
3. Wrong invoice number	Ⅰ Ⅰ	2
4. Pricing error	⊬⊬⊦ ⊬⊬⊦ ⊬⊬⊦ ⊬⊬⊦ ⊬⊬⊦ ⊬⊬⊦ ⊬⊬⊦ ⊬⊬⊦ Ⅰ Ⅰ Ⅰ	43
5. Wrong quantity indicated	⊬⊬⊦ Ⅰ Ⅰ	7
6. Wrong freight charge	⊬⊬⊦ ⊬⊬⊦ ⊬⊬⊦ ⊬⊬⊦ Ⅰ Ⅰ Ⅰ	23

Exhibit 3.7 Completed check sheet.

Check sheets are fundamental to data collection. They are the most common tools in TQS.

Pareto Charts

The Pareto principle comes from the nineteenth-century economist, Vilfredo Pareto, who found an uneven distribution of wealth. He noticed that 20 percent of the people owned 80 percent of the wealth. Juran found this uneven distribution was true in other areas as well; most of the trouble or opportunity comes from just a few causes. Thus, note how the **Pareto chart** in Exhibit 3.8 visually helps you separate the "vital few" from the "trivial many."

Assume that items A through F are the reasons why there were invoice errors in an insurance office: Item A was an incorrect claim number; Item B was an incomplete form; and so on. As you can see, items A and B account for nearly 70 percent of the invoice errors. Now you have clear direction on what needs to be investigated for correction. The initial causes may then be announced as incorrect claim numbers and incomplete forms.

A **quality improvement team** (QIT) went back and hypothesized why employees doing the invoices were not completing the forms and why the claim numbers were wrong. The QIT did some ad hoc

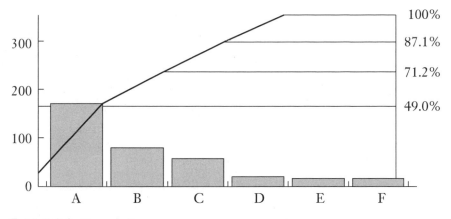

Exhibit 3.8   Pareto diagram.

research with those making the mistakes. The QIT found temporary employees doing the input. They hadn't received much training. Those filling out the forms were usually interrupted by calls and lost their focus. When the insurance company hired permanent employees for this function and trained them well, the claim number problems disappeared. When the employees were allowed to finish each claim form without interruption, the incomplete form problem disappeared.

Summary: Sample Sequence of Tools

Exhibit 3.9 shows how the tools may be combined. A customer problem emerges. The company then scopes the problem to find out where it is happening, how often, when, and by whom. A **quality improvement team** gets together to brainstorm possible causes of the problem and prioritize them with a **Pareto chart.** Key reasons for the problem emerge. The root causes are assessed in a **fishbone diagram.** The QIT reconvenes and brainstorms possible solutions. The solutions are tested, and they work. **Control charts** reveal that errors are now down. The QIT suggests improvements to other groups and sets standards for them. The QIT helps the organization celebrate the success and communicates it to both employees and clients.

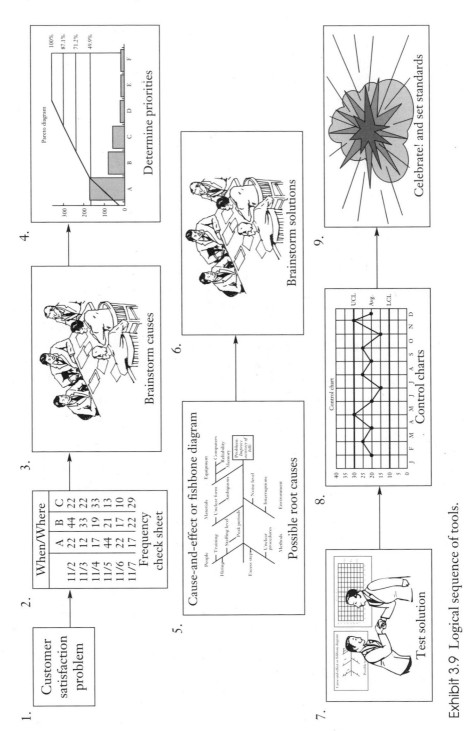

Exhibit 3.9 Logical sequence of tools.

3.0

Customer Focus and Satisfaction

Why Measure Customer Needs and Satisfaction?

Continuous improvement efforts in an organization must be anchored in customer needs and satisfaction. A frequent mistake in the United States has been to focus the continuous improvement effort on what managers or employees *assume* is important to customers. Managers will be the first to say, "We have contact with our customers all the time; we know what they want." I usually have managers then write down the top 10 needs of their customers. The overlap between the managers' list and that of the customers can be nominal.

Why? The data collection process behind the managers' assumptions was usually the squeaky wheel data collection system. The louder the squeaky wheel the more pervasive the managers assumed the need was. Being scientific and systematic are central ingredients to success. This is the one area in the Baldrige Award's seven criteria where having an outside consultant to help train you is imperative. If you aren't going to measure customer needs and satisfaction well, you really shouldn't bother. Since these data drive your whole quality process, having faulty data can lead you into irrelevant efforts. You can look very busy while going out of business.

Customer Needs Assessment

A **customer needs assessment** tells you which specific needs are or are not fulfilled by your product or service. One set of customers may want an automobile that has high status, reliability, a quiet and smooth ride, and hassle-free service. Thus, the Lexus. Another set may want a four-wheel drive utility vehicle that gets good gas mileage and is agile. Thus, the Jeep Eagle.

Customer needs cannot be easily assessed by written surveys. Most people don't like to write. Likewise, customers don't spend their waking hours thinking about the nuances of your service or product. Most have a hard time articulating their specific needs. Thus, a trained interviewer can help customers clearly define what specific needs are met by your service or product. Thus, the main tools for **customer needs assessment** center on

- **Focus groups** or **interviews**
- Customer advisory groups
- Observation
- Predictive measures of customer satisfaction and dissatisfaction (quality data)

Customer Focus and Satisfaction Measures

Customer focus and satisfaction measures assess whether the service or quality meets expectations. The formula for service quality is

Results – expectations = Service quality.

If someone expects extensive service and good food on a particular airline, the first-class cabin must deliver above that precise expectation. American Airlines determined that first-class customers liked their almonds warmed to around 99 degrees and that they liked meals served immediately on long flights. Customers pay extra for this service so they expect it. If people expect fewer services but good prices they may choose Southwest Airlines. Southwest has no full meal service and no baggage transfer. Yet it has consistently received the highest customer service ratings of all the airlines. Why? Because Southwest doesn't promise anything. Thus, when people board and find that they get an extra element of fun and friendliness, Southwest's ratings go up. Customers' perception of service quality is directly related to their experiences relative to expectations.

Customer focus and satisfaction measures are quite varied. They include both soft and hard measures. Soft measures center on customer perception. What people say and what they do may be different. An individual may be delighted with the taste of the extra raisins and nuts offered in the high-priced, well-advertised cereal but end up buying the generic brand. On the other hand, hard measures are behaviors and are measured in numbers like cash register receipts, market share, revenues, and profits. Both are important to understand customers. Soft measures help you understand what customers think of your service or product relative to your competition. Hard measures keep you informed about what they are actually buying.

Soft measures for customer satisfaction include surveys, **focus groups** or **interviews,** and observation. Hard measures include customer retention levels, market share, number of referrals from other customers, and revenue.

Caveats: Many managers starting in quality decide to throw together a quick customer satisfaction survey. Managers must do some homework before they send out such a survey. **Focus groups** or **interviews** are imperative so that the right questions can be covered in the survey. Otherwise, you can deploy your money on training and improvement on areas *that are not important to your customers.*

Tools, such as surveys, **focus groups**, and so on, need to be matched with the objectives. If you want good qualitative data (what questions to ask, what is important to people), you need a personal approach like **focus groups** and **interviews.** If you want to check out the data over a large number of people and get quantitative data, then written surveys are fine. Exhibit 3.10 helps you match your objectives to your tools. The additional costs involved in doing more personal measures will factor into a decision. Likewise, the timing of the data may be important. The customer cry of the 1990s is "Faster, Better, Cheaper." If your service or product changes quickly, then the type and frequency of the measure will also need to be considered.

Customer Satisfaction Surveys

Customer satisfaction surveys are written or telephone measures that determine levels of satisfaction with various facets of the product or service. Surveys are one of the most abused quality tools, especially for service companies where the relationship with the customer is vital. Many organizations quickly compose a superficial set of written questions and then send it to a sampling of their customers. Then

Objectives	Tool	Presentation
Show organizational comparisons	World-class surveys, across different companies using same survey; done by an independent researcher	Bar charts
	Benchmarking studies	
Identify key quality dimensions	**Focus groups** Interviews Observations	Bar charts
Monitor customer satisfaction over time	Feedback forms Observations Surveys	**Control charts** of overall satisfaction or key quality dimensions
		Either p chart, c chart, u chart, x chart, or s chart
		Percentage tracking on individual items
Develop or test ideas for services or products	Samples	Pie charts or bar charts
	Focus groups	Affinity diagrams
	QFD	QFD charts
Improve individual or team performance	Feedback forms	**Control charts**

Exhibit 3.10 Matching tools to measurement objectives.

these organizations base their entire quality strategies on a return rate of under 25 percent of the written questionnaires. The questions frequently don't represent what is important to customers and the answers are from too small a sample size. A more detailed treatment of how to design a real-time customer satisfaction measurement system is presented in my book *Measuring and Managing Customer Satisfaction: Going for the Gold*.

A better example is Eastman Chemical, a 1993 Baldrige Award winner. It thoroughly tested its survey before sending it out. Eastman's commitment to the results is evident in the way it deploys the survey. Salespeople hand deliver the surveys to over 18,000 worldwide customers. Then customers return the surveys to a neutral internal party. Customers are called if they have not returned the survey in a set amount of time. The salesperson then follows up with each customer and reviews the overall results and what is being done about each area indicated for correction. The return rate is over 70 percent.[25]

Surveys can provide a way to compare **key quality indicators** from year to year. Surveys provide important information in the perception of your quality or service. If done well, they can also show how you rate compared to your competition. If used well, surveys are an important part of the overall tool kit.

Tips on Customer Satisfaction Surveys

Sample. Samples need to be statistically representative of your customer base. You might want to stratify your customer base into categories. For instance, a tax software TQS company might have the following:

- Frequency of use: Frequent user, infrequent, nonuser
- Categories of use: Accountants, bookkeepers, end users
- Demographics: Geography, age, income, or other demographic information

The company could then randomly sample from within those subsets.

Bias. Biased questions that reflect what you want to hear won't give you much insight. The National Rifle Association sent out a questionnaire that asked, "Do you think your constitutional rights should be violated?" A waiter at a restaurant asked, "Do you want me to pass on any compliments to the chef?"

Bias can be very difficult to detect in reviewing your own work. Have someone else double-check you to make sure your questions are unbiased.

Ambiguity. Questions with ambiguity are interpreted differently by various people. One popular ambiguous question is, "How would you rate the professionalism of our staff?" Does professionalism mean how well the staff members dressed? How quickly they responded? How accurate their answer was? Again, having people discuss their meaning of your questions will help avert ambiguous questions.

User-Friendly. A user-friendly survey is short and looks as if it will take less than three minutes to complete. Lots of open-ended questions are a turnoff. A user-friendly survey is simple, easy to complete, and easily folded to put in the return mail. Two pages is maximum. Large postcards are ideal. Watch for visual clutter. If the survey is not user-friendly, you initiate a bias in who returns the survey. Only the highly committed will bother.

Simplicity. A complex question asks several things at the same time. Asking people to agree or disagree with this statement will cause confusion, "Our staff was both fast and friendly."

Frequency of Measuring Customer Satisfaction

A yearly interval is maximum. Beyond a year people don't remember their experience or feelings, or they may have moved. If you are going to measure at more frequent intervals than quarterly, use a personal approach. Customers typically don't like to fill out forms.

Exhibit 3.11 highlights the steps in measuring customer satisfaction. Telephone surveys are becoming increasingly popular as companies realize that it is almost as expensive to get a high response (more than 80 percent) from a written survey as is is to call their customers.

How to Formulate Questions

There are several choices in how to format questions. The following are the most popular.

Yes/no questions

The food was served warm.	❐ Yes	❐ No
The answers were accurate.	❐ Yes	❐ No

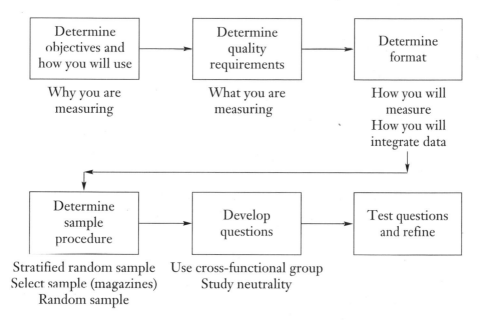

Exhibit 3.11 Steps in customer satisfaction measurement.

Likert scale questions

(The Likert scale is one of the least biased scales. Therefore, it is very popular.)

The food was served warm.

    ❒ Strongly agree ❒ Agree ❒ So-so ❒ Disagree ❒ Strongly disagree

The answers were accurate.

    ❒ Strongly agree ❒ Agree ❒ So-so ❒ Disagree ❒ Strongly disagree

The five-point Likert scale offers more accurate calibration of the answers. Having more than five points does not necessarily increase accuracy.[26]

Sample Customer Satisfaction Questionnaire

Note: The questionnaire in Exhibit 3.12 is meant to be an illustration, not one that can be lifted from this book and used with your customers. These questions are generic to service. Your questions need to be specific to your service or product. Likewise, you could be missing some vital questions that are critical to the buying behaviors of your

Survey

When was the last time you used this service or product?_____

How often have you used this service/product? At least:
❒ Once a day ❒ Once a week ❒ Once a month ❒ Once a year ❒ Never

What competitor products or services do you typically use? _____

Overall level of satisfaction:

1. What was your overall level of satisfaction with this service/product?
   ❒ Excellent ❒ Very good ❒ So-so ❒ Dissatisfied ❒ Strongly dissatisfied
2. I would use the service/product again.
   ❒ Strongly agree ❒ Agree ❒ So-so ❒ Disagree ❒ Strongly disagree
3. I would recommend this service/product to others.
   ❒ Strongly agree ❒ Agree ❒ So-so ❒ Disagree ❒ Strongly disagree

Circle the number that best represents your feelings. The first set of numbers relates to your expectations, the second to how the company performed.

	Degree of importance to me (1=unimportant to 5 = very important)	Degree of excellence that Company X does this (1= does not occur, 5 = always occurs)
1. Facilities are clean.	1 2 3 4 5	1 2 3 4 5
2. Employees are neat in appearance.	1 2 3 4 5	1 2 3 4 5
3. Employees respond in a timely manner.	1 2 3 4 5	1 2 3 4 5
4. Invoices are easy to understand.	1 2 3 4 5	1 2 3 4 5
5. Answers to questions are accurate.	1 2 3 4 5	1 2 3 4 5
6. Employees are courteous.	1 2 3 4 5	1 2 3 4 5
7. Problem solving is customer oriented.	1 2 3 4 5	1 2 3 4 5
8. Operating hours are convenient.	1 2 3 4 5	1 2 3 4 5
9. Manuals are easy to follow.	1 2 3 4 5	1 2 3 4 5
10. Customers receive personal attention.	1 2 3 4 5	1 2 3 4 5
11. Safety is emphasized.	1 2 3 4 5	1 2 3 4 5
12. Employees understand my needs.	1 2 3 4 5	1 2 3 4 5
13. The service is prompt.	1 2 3 4 5	1 2 3 4 5
14. Deadlines are met.	1 2 3 4 5	1 2 3 4 5
15. The product is easy to use.	1 2 3 4 5	1 2 3 4 5
16. The product is reliable.	1 2 3 4 5	1 2 3 4 5
17. The price is reasonable.	1 2 3 4 5	1 2 3 4 5
18. The location is convenient.	1 2 3 4 5	1 2 3 4 5
19. Forms are user-friendly.	1 2 3 4 5	1 2 3 4 5
20. The technology used is up-to-date.	1 2 3 4 5	1 2 3 4 5
21. The service/product is worth the price.	1 2 3 4 5	1 2 3 4 5

Name_____(optional) Phone number _____

Thanks for taking the time to help us serve you better! Please fold and return.

Exhibit 3.12 Sample customer satisfaction questionnaire.

customers. Thus, go through the extra step of finding out what questions *your customers* want to be asked.

Note in Exhibit 3.12 you will find a question that measures overall level of satisfaction. This question is the most highly correlated buying behavior. Include at least one overall question of satisfaction in your survey. The following are some examples.

1. What was your overall level of satisfaction with this service/product?

   ❐ Excellent ❐ Very good ❐ So-so ❐ Dissatisfied ❐ Strongly dissatisfied

2. I would use the service/product again.

   ❐ Strongly agree ❐ Agree ❐ So-so ❐ Disagree ❐ Strongly disagree

3. I would recommend this service/product to others.

   ❐ Strongly agree ❐ Agree ❐ So-so ❐ Disagree ❐ Strongly disagree

The first question on overall satisfaction offers the highest prediction of whether customers will buy your service or product again. Pay particular attention to the results of this question. The other question that is critical is question 21—the product or service is worth the price paid. AT&T has a measure called *customer value added (CVA)* that compares this perception of price with the perception of competitors' prices. AT&T found that the CVA measure is very highly correlated with buying behavior.[27]

Put your overall measures of satisfaction at the beginning of your survey. Overall satisfaction answers become distorted if they are preceded by lists of questions that begin to shape a person's thinking.

How to Tabulate and Show Results

Organizations can take the easy approach and just tabulate the percentage of people who responded favorably. Thus, the results might be shown in a graph like Exhibit 3.13

Tips on Surveys

1. Pilot your survey with representative customers to make sure that you have included the key questions and that the questions are clearly and neutrally worded.

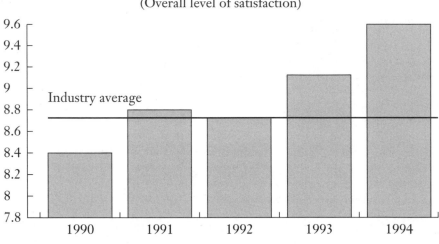

Exhibit 3.13 Customer satisfaction graph.

2. Be proactive about your surveys. Instead of sending them by mail, ask customers to fill them out before leaving your establishment. Create an incentive system for employees to have customers fill out the surveys. You can double your response rate just by having your employees ask your customers to fill out the survey and then asking customers if they have done so.

3. If you do mail the survey, include a motivating paragraph or letter. You need to encourage people to take the time and trouble to fill it out and send it in. Ways you can motivate people to complete the survey include the following:

    a. Tell customers how you will use the information to enhance your service to them.

    b. Let customers know their reactions are an important part of determining where you spend your improvement dollars.

    c. Add an incentive for returning the form. Incentives that work are extra frequent flyer points, a coupon for merchandise, or a chance at a drawing.

4. Ask for qualifying information. Questions should help you determine whether they are frequent customers or those who have never used your services.

5. Ask questions about your competition. Be sure to include your competitors' customers in your sample. This will give you good comparative data.

6. Leave room for comments. You might want two categories, such as what customers like and what suggestions they have for improvements.

7. Leave space for customers to sign their names and phone numbers. Specify that this information is optional. You then have a chance to call those people who responded in very positive or negative directions.

Key Quality Objectives

Use the customer satisfaction results to help extract those **key quality objectives** you want to improve. Exhibit 3.14 shows how to graph the results. These particular attributes came from a distribution TQS company, which found that late delivery and correct orders were two highly important areas to customers but where the distribution company performed poorly. These two areas became the priority focus.

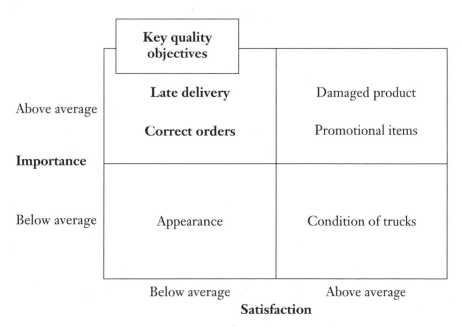

Exhibit 3.14 Choosing key quality objectives.

Focus Groups and Interviews

Focused interviews are one-on-one interviews that ask predesigned questions. They are frequently done by phone. **Focus groups** involve the same kind of queries but are done in groups of five to ten people. Usually a neutral, trained facilitator (frequently an outsider to the organization) structures the group.

Focus groups and **interviews** are best suited for in-depth analysis of product or service or new product or service development. Invitations are critical. They make individuals feel special and purposeful. They should be specific about what will happen. It should be easy for attendees to get to the **focus group.** Also, remember the following:

- Have five to ten people in the group. More than ten inhibits full and open discussion.
- Video or audiotape the group. The audio or video can be used in planning meetings to give credibility to the data. The tape(s) also can be used to strengthen training.
- Use a neutral outside facilitator. An insider usually is too vested in the outcome to foster full honesty.
- Take notes on a flip chart and post them around the room. The visibility of recording what people are saying reinforces their contributions.
- Include only people who have used the product or service. You might do a separate focus group with the uninitiated and ask separate questions. Giving the same weight to sophisticated users as nonusers can contaminate the results.
- Include people who have used competitors' products or services. Competitor customers are some of the best sources of determining what is really important to people.

Benefits

Both **focus interviews** and **groups** are extremely valuable vehicles for digging below the surface of expressed customer needs. The trained interviewer can establish a rapport that allows full honesty and disclosure. Likewise, a trained interviewer departs from preset questions and knows how to optimize the tangents the client may bring up. Since 80 percent of most innovations come from customers, these needs assessment devices allow for the expression of critical new ideas. Note that managers need to have their ears and eyes open for these innovative ideas from customers *at all times*, not just during needs assessment season.

How to Do a Focus Interview or Group

1. Hire or recruit a skilled facilitator.

2. Use a group process made up of representative company people to refine the questions that will be asked.

3. Decide on the place and time the **focus group(s)** will be held. Make sure both the time and place are convenient for participants. Trade shows are excellent opportunities to combine a **focus group** with a free dinner or appetizers.

4. Decide on incentives you will offer. Free food is a must. The food doesn't have to be elaborate. Drawings can also be helpful. Cash incentives are also helpful.

5. Decide who you will invite. You will probably want to do several **focus groups.** Consider asking persons who fit the following categories:
   a. Frequent users of your products or services
   b. Frequent users of your competitors' products or services
   c. Articulate and open individuals

6. Develop an invitation that will motivate people to come. Ways you can increase that desire are as follows:
   a. Tell them why they were invited; that is, why they are special.
   b. Work especially hard at recruiting a well-respected competitor in the field. Then drop that person's name to the others.
   c. Tell participants that there will be free food.
   d. Tell them the importance of their input to your improvement strategies.
   e. Make sure the time requirement is reasonable and the location easy for participants to access.

Sample Focus Group Questions

Exhibit 3.15 provides a sample outline of comments and questions in a **focus group.** This is just meant to be a guide. Specific questions need to be developed for each company doing this research, since the purpose of the **focus group** is to take advantage of the unique experience each person brings in the purchase and use of your product or service. Questions must be tailored to fit those specifics.

How to Use the Data

Audio or videotaping the sessions can be very helpful. You do need to ask permission of participants to record. Offer to turn the tape off if it will restrict honesty. Tell participants how you will use the tapes. The choices are to allow managers or employees to hear customer needs firsthand or to review areas that were not clear.

Use the input to generate hypotheses. **Focus groups** take time and money and usually don't represent the breadth or depth of your client database. Use the information to develop better survey questions and then test them over a wider group of people. Needs can be used in **quality function deployment**. Ideas for innovations or new products can be passed on to the appropriate departments.

Your thank-you note to focus interview or group participants might include changes that resulted from their discussion. Be sure your follow-up is timely.

Caveats: Many times managers new to **focus groups** will state a fear about combining their own customers from various organizations in the same room. Many of these customers may compete with one another. If you are selling accounting services to hospitals, for instance, those client hospitals may be competing with each other for patients. The managers might say, "Competitors won't want to get together and share in front of each other."

This apparent negative reaction can become a positive motivation. Competitors are curious about each other. A trained facilitator, though, is needed to encourage them to open up in front of each other so all can benefit from the experience. One of the major advances of total quality management or service is helping organizations that normally compete understand that they need to cooperate to a certain extent in order to compete better as a group. Evidence of this cooperation is seen daily on television commercials for raisins, diary products,

1. Introductions and purpose of the group
2. Critical incidents
    2.1 Think back to a positive experience you had using this product or service. Talk about what made it a positive experience. Then dig for specifics.
    2.2 Think back to a negative experience you had using this product or service. Talk about what made it a negative experience. Then dig for specifics.
3. Buying behavior
    3.1 What was important to you in selecting this product or purchase?
    3.2 Where do you usually buy this product or service?
    3.3 Rank order these factors (in order of importance): A, B, C, D, etc.
4. Features
    4.1 What features are desirable in this service or product?
    4.2 Rank order the following features in order of importance.
    4.3 Describe a positive experience in one of your more desirable features.
    4.4 Describe a negative experience in one of your more desirable features.
5. Quality
    5.1 What are the various issues in quality that are important about the service or product?
    5.2 Describe how we compare to others on those dimensions.
    5.3 What other suggestions do you have for improving our service/or product?
6. Pricing
    6.1 How important is price in you buying decision?
    6.2 Rate how you feel about the price we charge relative to the value received (1 = excellent to 5 = poor). Why?
    6.3 Rate Competitor A's product or service on the same scale.
    6.4 If we were to charge less, what features would you consider easy to give up?
    6.5 If we added feature A, what additional price would you be willing to pay?

Exhibit 3.15 Sample focus group questions.

meat, eggs, and so on. Competitors must come together to advance their industry as a whole.

Other Customer Satisfaction Measures

Surveys and focus interviews and groups are not the only measures of **customer focus and satisfaction.** Many organizations are concentrating on making the measurement system one that solves the problems on the spot. Thus, Baldrige Award–winning companies like Ritz-Carlton are focusing less on surveys and more on helping employees detect upset customers. As noted, Ritz-Carlton empowers employees to fix the problem on the spot. Employees have both permission to take time out from their regular duties and have a $2000 spending authority to fix problems. A fix-it feedback mechanism is much more powerful in customer retention than is a stale survey that arrives a year after the customer has defected to the competition. Keep in mind, 96 percent of customers never complain.[28] They just leave. Also, a customer who complains is likely to tell 10 other people about his or her dissatisfaction.

A Technical Assistance Research Program also found that 63 percent of all dissatisfied customers will never do business with an offending company again. If a supplier resolves its problems, 90 percent of those dissatisfied customers will remain loyal to the supplier. MBNA , the fourth largest bank credit card issuer, set up a special unit to ask customers why they defected. By fixing the problems on the spot, they retrieve about 50 percent of those customers who have just departed. MBNA keeps its customers twice as long as the industry average and MBNA customers run higher balances than other houses.[29]

Hewlett-Packard assigns an "owner" to every piece of customer feedback—both corporate and individual. This owner must act on the information and report back to the customer.[30] Standards are set for appropriate turnaround times.

Working with a complaining customer allows for instant reversals. Thus, examples of other customer feedback mechanisms are as follows:

- Fix-it feedback mechanism   Employees fix the problem as soon as they observe it. Incidents are documented.

- Transaction feedback forms

  After specified transactions (delivery, invoice, and so on), the customer is asked to fill out a brief card or answer questions.

- Observation

  Employees are taught how to observe nonverbal communication. Spa Resort Hawaiian, a Deming Prize winner in Japan, even uses videotapes to measure restlessness in the resort waiting lines.[31] Staffing levels were then determined to avert restlessness.

- Management phone surveys

  Many Baldrige Award–winning companies have managers assigned to call certain customers systematically and ask key quality questions.

- Customer appreciation days

  Many Marriott hotels have Wednesday night cocktail parties for their Platinum (highest level) Honored Guests. The staff mingles with guests and asks customer satisfaction questions.

- Customer advisory groups

  Regular meetings are set up with key customers to shape strategy and improvements.

- Customer speakers at meetings

  Customers are invited to staff meetings at the organization to talk about what is important to them.

- Complaint tracking

  Patterns in complaints are analyzed to see what systemic improvements are needed. Numbers are tracked to see if efforts are getting results.

How and When to Use the Data

All of these data-gathering devices need to be integrated at some point. Meetings need to be held at regular intervals to correlate the data from the various sources. As Marilyn Carlson of GTE Directories, and Joe Nacchio of AT&T Consumer Communications Services, both said at the 1995 Quest for Excellence Conference, "It wasn't until we started integrating customer satisfaction and quality data into our strategic planning process, that we started making significant gains."

Companies need to customize at what intervals they look at the data. Many high-tech companies, like Toshiba and Zytec, have daily quality meetings but they are short and focused on specific production problems. Wainwright customer service employees call each of their customers monthly to get a report card.[32] Other engineering companies I have worked with have slower change cycles, and the customers do not want to be polled more than twice a year. Let your customers help you determine the right interval. Even though the data may be reviewed at daily, weekly, or monthly intervals, my experience has shown that having managers review many data sources simultaneously encourages them to step back and get the big picture. Off site retreats encourage that level-of-quality thinking.

Internal Customer Focus and Satisfaction Measures

Internal customers should be treated the same way as external customers. Internal customers are those groups that serve one other. Word processing may serve engineering. The dispatching group may serve the nurses, doctors, and emergency medical technicians. When support groups within an organization respect their internal customers fully, you know you have arrived. Internal customers are thus treated with just as much respect as external customers.

Many of the same measurement devices that are useful for external customers are also useful for internal customers. For example, if an engineering department supports sales, the engineering group might want to do focused interviews to find out what is important to sales in evaluating engineering performance. Then a written or phone survey, done at periodic intervals, makes the measurement systematic and scientific. Complaint tracking, observation, customer appreciation days, and other qualitative measures can also be helpful in adding depth to customer understanding.

The data can be used for performance reviews, promotions, and award ceremonies. My book *Team Selling* explains the power of having teams and departments sell their accomplishments to each other. Microsoft uses this vehicle both to keep internal groups highly motivated and to educate the organization about accomplishments of other departments. An organization that believes in itself is one that shares this confidence with customers.

Control charts and progress reports on internal customers can be just as viable as those that monitor external customer progress. Bulletin boards, newsletters, and announcements at meetings are all appropriate vehicles for visibility. The importance of these measurements helps support groups stay customer focused. The latest trend to go outside for internal services, outsourcing, is also keeping support groups customer focused. Stephen T. McClellan, an analyst with Merrill Lynch, talked about the growing tidal wave of outsourcing in data processing.[33] He said the beneficiary of data processing outsourcing is Electronic Data Systems, the company founded by Ross Perot. The company had about $750 million in revenue in 1984 and grew to $7 billion by 1991.

Benchmark Tips

Benchmarking is a focused, systematic way of improving quality by studying others who are better than you, and then applying their experiences to your company. Robert Camp, formerly of Xerox, benchmarked L. L. Bean's Maine distribution center. He found that L. L. Bean could pick and pack goods three and a half times faster than Xerox could handle its spare parts. Xerox benchmarked John Deere for data processing.[34] Marriott looked at Milliken to determine employee involvement. Federal Express also looked at L. L. Bean to help figure out distribution. Southwest Airlines studied the Indy 500 pit crews to see how they serviced the cars so quickly. Benchmarking was the main tool used to create Saturn. The Saturn Project researched new ways for management and labor to work together while the Group of 99 benchmarked 60 successful companies to see how they operated.[35]

At Fluor Daniel, employees looked at both the recognized competitors and clients to see how they did major proposals. From studying how others organized support groups, trained and compensated their salespeople, wrote proposals, and rewarded those involved with proposals, Fluor Daniel changed its success rate from 23 percent to 76 percent on major campaigns. That represented billions of dollars in revenues for the Fortune 100 engineering company. Benchmarking others who do a function better than you can have big payoffs!

Who Does Benchmarking?

Some Baldrige Award winners have very formal benchmarking processes and some have quite informal processes. AT&T, Digital Equipment, DuPont, Eastman Kodak, and IBM have benchmarking departments that perform hundreds of studies.[36] Most teach employees how to benchmark to get the most out of the experience. Robert Camp, a former inventory control analyst at Xerox, designed a benchmark model that has been used by many companies. Small companies like Wainwright and Zytec encourage all their employees to benchmark. Zytec allows a travel budget to anyone who wants to visit a client or competitor organization.[37] Carl Arendt of the Westinghouse Productivity and Quality Center, says that he seldom finds a need for long, drawn-out studies. Sometimes a few calls will raise the necessary information.[38] This book is not a procedure book on the formal, regimented process. You might review Camp's book *Benchmarking: The Search for Industry Best Practices That Lead To Superior Performance*, for more details.

Neutral companies also offer **benchmark** services for satisfaction data among similar competitors. J. D. Powers offers **benchmark** data for the auto industry on customer perception of reliability and service. Trish, the power supply professional organization, allows organizations like Zytec to compare customer perception of its power supplies across its competitors.

In looking for areas to consider for benchmarking, consider the following questions.

- What is important to the customer?
- What is the value being added for the customer by benchmarking?
- How do we measure value?

- Who is doing the job well?
- What are the performance gaps?

Benefits of Benchmarking

Benchmarking *stimulates creativity.* Employees are forced to see that others may have even better ideas than the systems the employees produced.

Benchmarking *encourages honesty.* It's hard to think you're the best when you see a competitor or another client doing the same task with fewer errors and twice as fast. Even though access to competitors may not be easy, most of the industries I have worked with (engineering, publishing, computers, software, broadcasting, microchips, insurance, hotels, transportation, communications, and hospitals) have all participated in some form of sharing information with competitors in order to better their industries. The ethical rule of thumb is that benchmarking with competitors must be reciprocal. Don't ask unless you are willing to share too.

Benchmarking *heightens the improvement imperative.* When others set a better example, it creates a momentum that transfers to other operations. The Baldrige Award criteria uses the term *world-class* to reflect the stretch goal of being a world-class leader in your various processes. You only know that you are world-class through comparative data.

How to Benchmark

Some of the distilled steps in benchmarking are outlined in Michael J. Spendolini's book.[39] Those steps are as follows:

1. Carefully define the process to be benchmarked. Don't go in with sweeping questions like "How do you pay people?" A precise question might be "How do you use incentive systems with your salespeople to get high performance?"

2. Have a team do the benchmarking rather than an individual. The breadth of perspective and skills will improve the depth and quality of the questions. A team will also help convince others that a new practice or procedure may be worth considering.

3. Carefully select benchmarking team members. Select them on their ability to influence others as well as their relevant skill sets.

Credibility within the TQS company and a team spirit are also useful attributes.

4. Collect and analyze information. Benchmarking successes within your own organization will help you understand comparative data. It will also raise questions that may not be top-of-mind. Make sure the benchmarking phone call or visit has adequate preparation.

5. Adapt the benchmarked processes to your own institution and implement them. Direct transplants may not work because the organizations are different.

Other Benchmarking Tips

1. Plan your **benchmark** experience. Involve customers of the benchmarked data. Form a team and work flow the process. This will help generate detailed questions. Establish how you will measure the processes under investigation; for example, what metrics you will use. Plan the specific roles of team members. You might want to split forces at times and regroup for meetings at other times. Select people to **benchmark** who are curious, open-minded, and preferably articulate about what they find. Including a management sponsor is also important to implementing changes that may come out of the benchmark experience.

2. Be courteous to your **benchmark** target. It takes time and energy out of the partner's schedule. Offer to share information about yourself. Find out what the partner wants to know ahead of time.

3. Research your subject. Check internal experts and libraries to learn as much as you can about the company and process you are **benchmarking.** Mail or fax the questions in advance.

4. **Benchmark.** Use your checklists to make sure you cover the desired ground. Debrief immediately to consolidate the information.

5. Analyze the results of the **benchmark** with experts and managers. Determine the gap between your performance and your **benchmarking** partner. Determine steps to narrow the performance gap.

6. Develop an implementation plan to make the improvements. Assign roles, dates, and metrics for that improvement.

7. Announce the improvement to the rest of the organization. This fans the fires of both **benchmarking** and improvements.

4.0

Strategic Planning

The **strategic plan** is the overall road map for organizational success. Some companies separate strategic and quality plans and others combine them into one document. As long as the processes are done in conjunction with one another, it doesn't matter whether the documents are separate. Both need to be kept short and simple. Five to twelve pages is plenty. Backup documentation can be packaged separately.

When to Use Consultants

In the entire TQS road map, three critical processes determine a majority of your financial success. The first is whether your **steering committee** and leadership really understand TQS.[40] The second is the rigor of the **strategic planning** process. The third is the integrity of your **customer satisfaction measurement** system. Bain and Company, a management consulting firm, studied 463 companies and analyzed which of the 25 standard quality tools they were using (mission statements, SPC, customer satisfaction measures, and so on). Bain and Company then correlated the tools with financial performance. Customer satisfaction measurement had the highest relationship to financial success.[41]

If you only had three places to get help from the outside, these would be the three. Many companies use outside help to train their **steering committee,** facilitate in **strategic planning,** and set up a **customer satisfaction** system. Many organizations spend all their consulting money on training employees on concepts and tools. That can be done by insiders. Outsiders can provide both rigor, expertise, political safety, and new views in the three critical areas of TQS.

In **strategic planning,** the essential ingredients are as follows:

- **State-of-the-organization report** This short report gives a snapshot of the current organization. It profiles key clients, most profitable products/services, highest growth areas, the financial picture, and employee inventory. Most companies need to look at the data in order to prevent being misled by their own assumptions.

- **Strategic analysis and plan** The analysis provides needed homework in order to make intelligent decisions about new directions in marketing, R & D, target customers, and changes in product/service lines. Without doing a **strategic analysis,** the **strategic plan** is little more than guesswork.

  The **strategic plan** presents the conclusions that arise from going through the strategic analysis and **state-of-the-organization report.** New goals are set in areas of market share, target clients, target products/services, and target areas for R & D. The plan also details how the goals will be accomplished and measured.

- **Quality analysis and plan** Quality analysis takes a look at the **key processes** in the organization and decides on the **key quality indicators.** The **quality plan** stipulates which areas will be the **key quality objectives** for the year. How those objectives will be achieved is also spelled out.

Intervals. A major **strategic plan** is done either at three-, five-, or ten-year intervals, depending on how quickly your service or products change. An engineering company can get by with a five-year major effort. High tech moves quickly enough that a long-term plan every five years would mean that an organization could be out of business. Long-term goals, however, are critical for everyone. Motorola and Hewlett-Packard set goals to increase their defect-free rate tenfold over ten years. Both achieved this ambitious goal. Each year, however, the specific objectives need to be updated.

Deployment. The planning process is just as important as the plan. The executive committee normally comes out with a draft plan and has low-level managers provide input. Each level down becomes progressively specific about how its division or department plans to achieve goals within the plan.

Many Deming Prize– and Baldrige Award–winning companies (Toyota, Komatsu, Juki, Zytec, and so on) give each employee a pocket-sized book that contains the following:

1. Mission statement of the company
2. State-of-the-organization capsule (key clients, competitors, and products)
3. Strategic quality objective(s) for that year for the company
4. Strategic quality objectives for that year for the division and/or department
5. Review of quality tools (statistical process control, flowcharts, and so on)

The next two pages of the small book are blank to allow the company to insert self-adhesive pages every year as objectives are revised. This pocket-sized quality book is the basis for all performance reviews, promotions, and hiring objectives, and is the basic direction of the TQS company.

Planning is 20 percent of the continuous improvement process, and implementation is 80 percent. Review both the leadership and accountability systems sections of this book to see how follow-up is done. Keep in mind that upper-level managers in high-performing quality companies spend 30 percent to 50 percent of their time planning progress, reviewing results, removing obstacles for employees to achieve results, and celebrating progress.

Caveats: The **strategic plan** varies in size from three to twelve pages. Save the lengthy analysis for a separate set of papers. More than 12 pages won't be read by the employees, clients, suppliers, and others that need to be involved. Quantitative information is best done in graphic form.

Strategic Analysis

Strategic analysis includes all the homework you do on your changing competitors, clients, industry trends, and your organization's strengths and weaknesses. It is needed to do a **strategic plan** based on data rather than guesswork. Small companies can probably do a worthwhile analysis in a few days. Large companies may take months.

Components

Industry Trend Analysis. Unless you focus on regulatory, value, financial, and global changes, you can find yourself making slide rules in an age of calculators. This was both IBM and Digital Equipment's fatal flaw during the late 1980s.[42] They refused to pay attention to the fact that business was switching from mainframe computers to distributed personal computers. The anchor stores at shopping malls refused to notice that customers wanted more say in buying and service delivery. Robinsons, Bullocks, Buffums, Bloomingdales, and others all hit major financial disasters as stores like the Limited, Victoria's Secret, and Lerner grew as they responded more to customer needs.

Advertising and brand loyalty in the 1980s were replaced by discounting and specialization in the 1990s. The mergers and Wall Street focus on making money that drove the 1980s left people insensitive to the effects of their buying decisions in areas like the environment, violence, family values, and other fundamentals. In a strategic analysis I did for a major client in 1993, my colleagues and I conducted 65 focused interviews and groups for retailers in five medium-sized cities. We found that the 1990s values had shifted back to basics. Mall marketers reported that their customers now wanted a safe, clean, fun shopping experience. Unless the malls were attentive to those needs, they saw their customers disappear into the discount stores like Costco, Price Club, and Sam's Club.

Government regulations change fundamental structures. In the mid-1980s hospitals began an intense focus on increasing government regulations. Some hospital administrators saw it as a crisis and responded by becoming totally lost in the detail of dozens of new regulations. Other visionaries developed a strategy and quality plan and positioned themselves against what was to become fierce competition for survival. Hospitals in Rochester, New York and Portland, Oregon have set examples in how constraints can be combined with expansion. Systematic review of outside changes can prevent an organization from being ambushed by what others are using to their advantage.

Competitor Profiles. These help you update your knowledge of your competition. When the environmental division and the telecommunications division at Fluor Daniel were started, employees did extensive review of the strengths and weaknesses of current competitors. When a strategic realignment was done during the mid-1980s, each new market-focused division was responsible for developing its own **strategic plan.** Part of this plan included analyzing competitors and client needs and determining the best strategy. This profiling is best done in the major **strategic plan** and then updated yearly. Use the **benchmarking** processes to gain detailed insight into the competitors' processes.

Strategic Synthesis. This synthesis combines what you have gleaned from your **focus groups**, competitor profiles, and self-analysis. Use it to establish your key selling points—those areas that your clients need where you are clearly ahead of the competition. Make these key selling points known to all of your employees, suppliers, and clients. The strategy is synthesized in many mission and vision statements. At Fluor Daniel, the mission was to, *Enhance the competitive advantage of our customers by providing quality services at unmatched prices.*

That mission statement was translated into specific strategies that would enhance the company's ability to provide its customers with competitive advantage. Fluor Daniel would focus on hiring the top-notch experts in clean room engineering to build production facilities for pharmaceutical plants. Other efforts focused on getting costs in line to provide "unmatched value," which was a measurable goal.

In more spirited companies, the positioning statement can also turn into a battle cry. Many of the quarterly meetings held when I worked at Microsoft engaged the spirit of the employees through fun expressions of these battle cries. Bill Gates would come riding on to the stage on a white horse with swirls of white fog at his feet while we sang the battle cry. We would do skits that portrayed Microsoft's competitive edge. We turned the strategy into fun meetings that would test our understanding of the implementation plan and engage the competitive spirit. Wal Mart has similar pep rallies. Ben and Jerry's has taken the quality-of-life part of its mission to new heights in its incorporation of "fun as a corporate mantra." The Joy Gang is responsible for spreading merrymaking.[43] The point in these examples is that what is considered hoopla by some, captures the heart and souls of others in organizations that engage the spirit of their employees. Too many companies forget to have fun on their way to their strategic goals.

Deployment

Responsibilities for this analysis need to be spread over the executive committee or quality **steering committee.** Never should it be done by one staff member. One person needs to be responsible for giving directions, setting page limitations, and synthesizing results. The main benefit is derived from many people doing the homework. The benchmarking data that are ongoing may be relevant to pull into the strategic analysis.

One tool that has been helpful to many companies has been a way of capturing employees' talents so they can access each other's expertise. This is done by a topical listing of who within the company is an expert on what. These "Yellow Pages" showcase those experts on the company's competitors, clients, products, services, or processes. A medical software company may have resident experts on Microsoft, Aldus, IBM, Apple, Compaq, and so on. The same company may have identified experts on different types of networks, such as Novell and Gateway. Likewise, the company might pinpoint experts on its specific clients, such as Humana and Hoag, as well as on its competitors. Responsibilities of the experts include reading the trade magazines for that specialty or company, going to professional meetings, and reporting back any changes.

A well-developed Yellow Pages system allows you to instantly push the organizational buttons to do a **strategic analysis.** People in the Yellow Pages also become resources for others who may be working on proposals, joint venture work, new services, or products. Yellow Pages prevent a fairly common occurrence within many large organizations. A group doing a new product or service launch, or writing a proposal, can save considerable original research when instant experts are readily accessible. Yellow Pages provide a quick path to the organizational pockets of gold. The Yellow Pages have been useful in the many companies with which I have worked.

State-of-the-Organization Report

The **state-of-the-organization report** is a self-analysis of the current facts of the organization. Exhibit 3.16 provides a template for some of the questions that are important to answer in that self-analysis. This simplified form is intended for small- to medium-sized businesses. Large companies usually have elaborate self-analysis. Many companies get out of touch with their shifting profit centers and key clients. This report helps an organization obtain a snapshot or current view.

Name _____   Company _____

Address _____

Phone _____   Position _____

	Where	Number of employees
Headquarters	_____	_____
Other major offices	_____	_____
	_____	_____
	_____	_____

What is the company's mission? _____
_____

What are the top-selling products/services your company produces and what market share does each enjoy?

	Products/services	Market share (%)
1.	_____	_____
2.	_____	_____
3.	_____	_____

What were the approximate revenues from last year? _____
What percentage represents your cost of sales?   _____
What percentage of revenues were your profits?   _____

What is the recent financial history of the company?
Growth _____   Steady state _____   Downsizing _____

How does your company spend its money? Draw a pie chart of outlay according to categories such as salaries and benefits, production/manufacturing expenses, training, sales, equipment, R & D, and so on.

Example

Salaries and benefits 40%
Expenses and overhead 23%
Training 7%
Equipment 20%
R&D 10%

Your company's output

Exhibit 3.16 State-of-the-organization questionnaire.

What are the trends influencing your sales (economic, legislative, demographic, and so on)?

1. _____
2. _____
3. _____

Who are your key competitors? Their market share? Strengths? Weaknesses?

	Competitor	Market share (%)	Strengths	Weaknesses
1.	_____	_____	_____	_____
2.	_____	_____	_____	_____
3.	_____	_____	_____	_____

What value do you feel your service/product has over the competition?
_____
_____

How is your company currently measuring customer focus and satisfaction?
_____
_____

What client tracking systems (computer or otherwise) do you currently have?

1. _____
2. _____
3. _____
4. _____

Describe any marketing or sales planning system you are currently using.
_____
_____

Is your company implementing a total quality service or total quality management program?   ❒ Yes   ❒ No

If yes, what involvement has sales and marketing had in that program?
_____
_____

Exhibit 3.16 *(continued)*

What specific areas would you like to see the company (and/or sales and marketing area) improve on?

1. _____
2. _____
3. _____

Who are your key clients?

1. _____
2. _____
3. _____
4. _____
5. _____
6. _____

Exhibit 3.16 *(continued)*

Industry Trend Analysis

The **industry trend analysis** helps your organization keep abreast of regulatory, value, financial, and other changes that impact your company. Paragraph titles need to include the following:

- Regulatory changes impacting the organization
- Value changes likely to impact your services/products
- Relevant economic trends
- Relevant global changes
- Relevant buying preference changes
- Relevant demographic changes (age, sex, personalities, locations)
- Relevant technological changes

Competitor Profile

Competitor profiles detail essential facts about your competitors. **Competitor profiles** can be used in preparation for benchmarking efforts, for proposals, and for a better understanding of how to set competitive edges. Exhibit 3.17 provides an example.

_____
(name of competitor)

1. Location   Where   Number of employees
   1.1 Headquarters location _____   _____
   1.2 Subsidiary locations   _____   _____

2. Finance   1994   1995   1996
   Annual revenue   _____   _____   _____
   Earnings   _____   _____   _____
   Return on investment _____   _____   _____
   Return on equity   _____   _____   _____
   Price/earnings ratio _____   _____   _____

3. List the company's top three target areas and the market share in
   each.
          Target area   % of their business   Market share
   3.1 _____   _____   _____
   3.2 _____   _____   _____
   3.3 _____   _____   _____

4. For what is the company is best known?
   _____
   _____
   _____

5. What growth areas are targeted by this company?
   _____
   _____
   _____

6. Rate the reputation of the company in its key areas of
   performance (1 = low to 5 = high).
          Key area   Reputation
   6.1 _____   _____
   6.2 _____   _____
   6.3 _____   _____

7. What major sales has the company had in the past two years?
   _____
   _____

Exhibit 3.17 Competitor profile.

8. How centralized is the company?
_____
_____
_____

9. What is the company's history in the following areas (high, medium, low)?
   9.1 Employee turnover   _____
   9.2 Company's benefits   _____
   9.3 Employee morale   _____
   9.4 Use of subcontractors   _____
   9.5 Cost of sales   _____
   9.6 Client satisfaction   _____

10. What supplier and client strategic alliances are in place?
_____
_____

11. Use this page to list questions that are specific to your industry. Include questions that help you compare the following:
    a. Key features of their products or services and yours
    b. How they organize their sales force
    c. Strengths and weaknesses of their organization, manufacturing, and so on.

Exhibit 3.17   *(continued)*

Strategic Synthesis

A **strategic synthesis** draws together information about client needs, your competitors' weaknesses, and your strengths. Exhibit 3.18 shows how the center ring is where your key selling points or central strategy come from.

Quality Analysis

Quality analysis is the process of identifying which processes will most likely benefit the customers and thus the bottom line. Identifying **key processes** and determining **key quality objectives** are the very core of

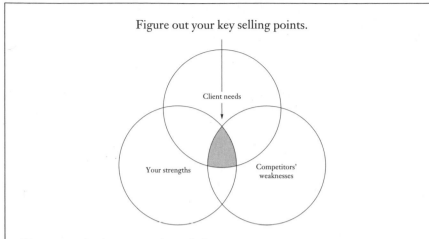

Figure out your key selling points.

How to synthesize a strategic analysis.
1. Determine client needs (from **focus groups,** surveys, and interviews).
2. Weight client needs (High, medium, low).
3. Compare your organization's strengths against your competitors on only the high client needs. Use a scale of 1–5 to compare (1 = low, 5 = high).
4. Pick out two to four important client needs where you excel and your competitors don't. Those are your key selling points.

Client needs	Importance	You	Competitors		
			1	2	3
1.					
2.					
3.					
4.					
5.					
6.					
7.					
8.					

Put an asterisk by the high client needs in which you are strong and your competitors are weak. Those are your key selling points. List them here.

1.   3.
2.   4.

Positioning statement
Synthesize your key selling points into one battle cry.
(Examples: Grossmont College Center for Vocationally Handicapped: Learn to earn; Wichita KKRD Radio Station: The #1 Hit Radio Station in Wichita)
Your positioning statement: _____
_____

Exhibit 3.18 Key selling points.

quality. Then focusing on implementation and employee involvement will make the difference in whether any real changes emerge.

Steps in Quality Analysis

1. Review the results from the **customer needs assessment** measures (client **focus groups,** surveys, complaint systems).
2. Review the areas where the competitors are better than you.
3. Identify the **key processes.**
4. Review your **strategic analysis** and **plan.**
5. Determine which three to five quality initiatives are key to your success.

Process

1. This is best done by an executive team or quality **steering committee.**
2. Invite key clients to the brainstorming session.
3. Pay particular attention to the data that your customers generate.
4. After you have decided what will be your **key quality objectives** for the year, announce them to all employees, suppliers, and clients.
5. Have each division, department, and team take those objectives and determine their own quality implementation plan.

Key Processes

As people learn about quality, they frequently ask, What is the difference between a system and a process? A system has a purpose. A human being is a system. Each individual determines his or her own purpose, from survival to changing the world. A tiger is a system. The tiger's purpose may be survival or it may be to entertain in a circus. The purpose of an organization is usually encapsulated in the mission statement. For instance, Disneyland's mission is "to create magic."

A system can be made up of many different processes. A process is a systematic sequence of actions necessary to accomplish some end.

Each person is also made up of many different processes; for instance, a circulatory, digestive, or neurological process. Each process has a sequence of steps that accomplishes some goal. Some of those processes are more vital to survival than others. If your digestive system develops a problem and creates too much acid, you get ulcers. You can be treated for ulcers. If your heart stops and does not circulate blood, you die. Thus, some aspects of the circulatory process have more immediate consequences for your survival than the digestive tract.

Every organization is a compilation of processes that allow it to take input, process that input, and turn it into a value-added service or product. The mission of the organization clearly expresses its purpose. Managers, then, need to look at what are the processes that go into the survival of the organization—the vital few. In education, the vital processes may be recruiting and teaching. In computer software, the vital processes may be new product development and marketing. In a hospital, the vital processes may be partnering with quality doctors and insurance. You determine vital processes through self-analysis *and* customer needs assessment. Vital processes need to reflect why customers choose you and why they might leave.

Exhibit 3.19 illustrates sample **key processes** of various organizations. These key processes are often called the critical competencies of the organization.

Eastman Chemical, a 1993 Baldrige Award winner, included in its award application 19 key processes that are managed by its executive committee.[44] Those processes transcend the organization and include **strategic planning**, communication, research and development, and

Colleges	Insurance	Software	Engineering
Recruiting	Claims	R & D	Disciplines
Enrollment	Actuarial	Programming	Projects
Orientation	Investment	Systems	Legal
Teaching	Sales	Supplier liaison	Cost
Financial aid	Marketing	Manuals	Scheduling
Registration	Suppliers	Production	Procurement
Faculty recruiting	Invoicing	Sales	Training
Grants and funding	Supplier liaison	Marketing	Quality control
		Quality control	Project management

Exhibit 3.19 Sample key processes.

others. Each of those processes has an improvement committee that sets goals and is responsible for meeting those goals. Performance reviews on team members are based on whether or not those goals have been reached.

How to Determine the Key Processes

1. Have the **executive team** brainstorm all of the organizational processes.
2. Have the team vote on which ones are key to **customer focus and satisfaction.**
3. Have the customers vote on the same thing.
4. Determine which are the **key processes** by voting on them.
5. The **key quality indicators** should fall within these **key processes.**
6. Determine the **key quality objectives** for each (see next section).

Example

An insurance company decided that the claims process was one of the most critical to customer satisfaction. The company had a three-month turnaround time between hearing about an auto accident and reimbursing the owner for damages. The managers decided to set a goal (**key quality initiative**) of three weeks for cycle time. They not only achieved the three-week turnaround time within a period of six months but they also went on to get it down to one day. Positive word-of-mouth advertising came from their customers.

Caveats: Make sure that you have done **focus groups** with customers to get their input. One of my clients, a bank, had focused its efforts on training the tellers to shorten the waiting line in the bank. If the managers had done their homework, however, they would have found that over 20 percent of their customers had recently changed banks because their automatic teller machine (ATM) was in a dimly lit area. Security-conscious customers had switched to a bank that positioned its ATM for safety and had it well illuminated at night. Later this first bank found out that only 15 percent of its customers even came into the bank. Thus, this first bank could easily lose 85 percent of its customers while its managers focused on a lesser-important variable. The ATM turned out to be a **key process.** Increased security (or a perception of security) should have been a **key quality objective.**

A second warning needs to accompany the first. Many companies try to save money and use other companies' data on customers' needs. Beware! Security may be the number-one issue in Compton, California but the number-thirty issue in Sequim, Washington. Each TQS company needs to do local research as well as use broad data specific to its industry. Likewise, needs change over time. Determine at what intervals you want to update your data.

Key Quality Objectives

Key quality objectives (KQO) are key selling points on which you choose to focus your improvement effort. A **key quality objective** may be to reduce defects in your product by tenfold in ten years (Motorola). Another may be to reduce employee turnover. **Key quality objectives** are listed in your **strategic plan** and spell out your targets for improvement. You determine these through **focus groups** or **interviews.** Never just rely on managers to guess what they are. Also set goals that are quantum leaps ahead for your industry.

Check your key selling points. These should be the same or very close to your **key quality objectives.** If not, you need to have a good reason why not; for example, you may already be a world-class leader and need to work on some other aspects of your business. These **key quality indicators** are the basis on which you base your recruitment, promotion, reward, and other systems. More than four overall **key quality indicators** usually means a lack of focus for your organization. Exhibit 3.20 provides a sample of how to translate the key processes into **key quality indicators,** and finally, into specific **key quality objectives.**

Key process	Key quality indicators	Key quality objectives
Delivery of service	Cycle time from order to receipt	Three days
Design	Durability of product	10,000 hours mean time between failure (MTBF)
Billing	Billing accuracy	Six sigma (three defects per million)

Exhibit 3.20 Example of distilling key quality objectives.

Implementation

Key quality objectives are set at high levels and deployed throughout the ranks. Departmental, team, and individual objectives should all be able to contribute to these initiatives. Exhibit 3.21 indicates a way of spelling out the deployment of those objectives. It carries the **key quality objectives** into the project management phase of how goals will be measured, by whom, and when.

Key quality objectives	Goals for improvement	Measures	Who	When
1.				
2.				
3.				

Exhibit 3.21 Key quality objectives.

Types of Key Quality Indicators

Exhibit 3.22 provides examples of typical **key quality indicators** in a service organization. The indicators are divided into hard measures, such as those that relate to actual buying behavior, and soft measures, such as those that relate to perception.

Caveats: Three to five **key quality objectives** is plenty. Otherwise the employees lose focus and don't accomplish any of the objectives. As the objectives are deployed downward, each level becomes increasingly specific. Exhibit 3.23 demonstrates how the **key quality objectives** at the top are more general than the ones at the low levels of the organization.

Strategic Plan

The **strategic plan** is a summary of all the preceding work done in the strategic and quality analysis. The best plans are short on paperwork and long on implementation. A table of contents illustrates how the **strategic plan** merely assembles what you have done to date. Long-

Hard measures (Behavior-based data)	Soft measures (Perception-based data)
• Client defections • Employee turnover • Market share • Number of units sold or attendance • Percentage of repeat customers • Delivery cycle time • Defects • Price	• Client satisfaction measures • Employee satisfaction measures • Perception of service/product • Supplier satisfaction measures • Client perceptions relative to competitors • Complaints • Lost customer perceptions

Exhibit 3.22 Types of key quality indicators.

Level in company	Key quality objective
Company KQO	Improve cycle time from order to delivery.
Division KQO	Improve average cycle time on X service from three weeks to three days.
Department KQO (order entry group)	Enter orders within one hour of receipt.

Exhibit 3.23 Key quality objective deployment.

term **strategic plans** are usually done every three to five years and then updated every year. Exhibit 3.24 details a sample table of contents for a **strategic plan**.

Quality Control Plan

Quality control preceded the focus on upstream quality. Inspection was the early emphasis. Dr. W. Edwards Deming, at a seminar in San Jose, California, on July 21–22, 1992, talked about the historical roots of inspectors. During the 1940s, Western Electric employed over 40,000 people at its Hawthorne plant near Chicago. Over 25 percent of Hawthorne employees were inspectors. The **quality control plan**

**Strategic plan
Contents**

Brief history of company

Mission of company

State-of-the-organization report
State-of-the-organization questionnaire
Customer profiles
Competitor profiles
Trends and conclusions

Strategic plan (Analysis, Goals, and Implementation)
Strategic analysis and synthesis
Strategic goals
Strategic implementation plan

Quality plan (Analysis, Goals, and Implementation)
Quality analysis—summary of analysis done

Quality goals
Key quality indicators (initial)

Quality implementation
Steering committee description
Project plan for quality (from leadership section)
Systems plans (from leadership section)
Annual quality report (covers key quality objectives)
Quality review process (monthly, quarterly, who, how)
Report development process (who, when, what)

Exhibit 3.24 Sample table of contents.

laid out who was going to inspect whom and when. As the science of quality has evolved, the emphasis has shifted to prevention. Thus the **strategic plan.** It focuses on embedding continuous improvement and innovation throughout the organization. It sets goals on what specific areas are targeted for improvement for the year and lays out an implementation plan. The **strategic plan** is inspired by the Baldrige Award. **Quality control plans** are still required by the ISO 9000 standards and still require independent systems audits and self-inspections. Either internal or external assessments can still be helpful to systematic self-evaluation.

Most all major companies in the United States have **quality control plans** and quality inspectors. The trend is to downsize inspection departments and make self-inspection resident in all the workers. Because **quality control plans** are both old news and well evolved within most organizations, this book will not cover quality control. The ISO 9000 series standards cover issues related to quality control. Becoming certified to ISO 9000 is a major concern of many service companies in the 1990s, as they discover that this "badge" helps them do business in Europe.

Procedures Manual

The ISO 9000 series standards require that procedures be documented. Beyond that, procedures make it easy for standardization of performance within a company. The thought behind having procedures is that when someone leaves, it should be easy for the next person to pick up exactly where the first person left off. Procedures also help the organization consistently produce the same standard of quality.

High-quality companies all have standards and procedures. Take Disneyland's theme, "We create magic." In order to create a fun environment for guests, Disneyland regulates the dress, hair length, facial hair, and other appearance factors of its employees. Appearance and attitude are an important part of creating magic. Marriott has exacting volumes of specifications it requires for its hotel builders. When you stay at a Marriott, you know where the towel rack will be located, how much space will be in each room, and what basic colors will be used for interior decorating. Procedures are also exacting for employees, and deviations are not tolerated. Mavericks don't survive well with either Disney or Marriott. What emerges, though, is a consistent experience for guests of both establishments. Disneyland procedures are so well respected globally that the Japanese are studying how Disneyland has written and enforced its standards.[45]

How to Write Procedures

The ISO 9000 standards concentrate mainly on the **quality control plan** and procedures. If you plan to eventually apply for certification, I recommend you get outside training in how to write procedures. You can be overly detailed in your procedures and end up wasting money and time. If you are overly specific you can almost guarantee that the ISO 9000 auditors will find inconsistencies and errors. They are auditors and have a keen eye for such. Likewise, if you write your procedures from an overly general perspective, you will not meet the basic requirements. Procedures are meant to make workers interchangeable. If someone leaves, another individual should be able to pick up the well-written procedures for that job and provide quality output.

Example of Level of Detail Required

Too general: This job function involves calling on clients, writing proposals, setting the price for bids, and training new inside salespeople.

Too specific: The procedure for calling on a client includes the following:
1. Refer to the salesperson's daily calendar of appointments.
2. Call the client to confirm the appointment.
3. Walk out to the car.
4. Turn the key to the "on" position in the car with a foot on the brake.

Just right: The procedure for calling on a client is as follows:
1. Confirm the appointment.
2. Review the client file and plan the interview.
3. Call on the client.
4. Submit a visit report.

Sample Packaging Procedures from an ISO 9000– Registered Company

1. Overall: Unpacked finished goods delivered to shipping shall be packaged within 24 hours. Supplies shall be packed according to the work instructions defined in the "Directory of Packaging Instructions."

2. Storage: Finished goods shall be stored in an organized, secure manner to protect them against loss, damage, or deterioration.

3. Identification: Work order tickets shall be attached to products for identification purposes.

4. Inventory: A daily inventory and inspection of finished goods shall be performed by a traffic clerk.

5. Security: The shipping area shall be kept secure by keeping exterior doors locked at all times.

5.0

Human Resource Development and Management

Deployment of quality principles throughout employee ranks is the primary focus of the quality human resource (HR) and management process. According to the Baldrige Award criteria, the key elements of this area are

- **HR planning** and evaluation
- High-performance work systems
- **Employee education** and **plan training**
- **Employee well-being and satisfaction**

Employee performance reviews and recognition were covered under **quality accountability system** and **celebration systems** in the leadership section. The other categories will be covered in this section.

Baldrige Award examiners look for distributed power and knowledge among employees at all levels. In order for continuous improvement to work well, upper-level managers must respect employees just as much as they do their outside customers. Low-level employees need to understand the strategic direction of the company and the quality results that are applicable to their area.

AT&T Universal Card Services discovered that it needed to "delight" employees before employees could "delight" customers. Its HR systems were created by this principle. AT&T Universal Card Services looks for the best people in each job function. Then AT&T empowers employees to control their own destinies. AT&T Universal Card Services not only won the Baldrige Award but is also one of the most profitable credit card companies in the United States.

Training is the centerpiece of this section on HR development. U.S. companies spend $60 billion per year on training.[46] Over two-thirds of that money goes toward supervisory or management training. The Japanese spend a bit more, but two-thirds of their training dollars go to frontline worker training.[47] Frontline employees not only deliver the goods in service organizations, they are the goods. This section discusses how to invest in training dollars to develop these precious assets. As Jan Carlzon said of Scandinavian Airlines' service quality turnaround, "The assets side of our balance sheet should not cover the aircraft we have. It should list the numbers of satisfied customers we have—the number of passengers who would return to SAS. These customers are real assets." Employees need more than "smile training" to use the computer systems swiftly, deliver timely answers, and be able to make decisions.

HR Planning and Management

Human resource planning is an ingredient of the **strategic plan**. The implementation of the **strategic plan** should include a section under HR for both strategy and quality. Thus, the strategic goals state new targets for types of clients, products, services, or suppliers. The **key quality objectives** spell out long- and short-term quality initiatives. Those are handed to the various departments to iterate their own plans. The HR plan comes from that document and talks about how those goals will be implemented. Exhibit 3.25 provides a sample table of contents.

Employee Involvement Process

Getting employees involved in quality is a complicated effort. It is driven from the top by constant setting of strategic quality objectives and by attention to employee involvement efforts. It is driven from the bottom by employee initiatives and responsive managers. It is driven from the middle by **champions** who continuously solicit and repeat **success stories.**

Some companies mandate employee involvement and put all their employees on a team. That is a waste of both time and human spirit. Nothing is more demoralizing than going to meaningless meetings.

**Quality HR plan
Contents**

1. Reiteration of strategic and quality goal

2. Implementation of those goals
    2.1   Education, training, and development required
    2.2   Changes needed in the workforce to address mobility
           and cross-functional training
    2.3   Rewards, recognition, benefits, and compensation for
           accomplishing goals
    2.4   Recruitment—with additional focus on changes in the
           diversity of the workforce

3. How HR improves its own processes
    3.1.   Recruitment
    3.2   Training
    3.3   Rewards

4. How data are used to improve the overall working environment
    for employees
    4.1   How employee survey data are used
    4.2   How employees contribute to the **strategic plan**

Exhibit 3.25 Sample table of contents.

Managers are much better off leading by example. Use the **steering committee** to build teamwork and enthusiasm. Use the **champions** to start selling the process. Look for opportunities to put teams together and solve departmental or interdepartmental issues.

There are typically two types of motivation for employees to get involved: top-down and bottom-up.

Top-Down Employee Involvement

Top-down employee involvement is generated by the quality **steering committee** and the **strategic plan**. Strategic and long-term quality objectives require improvement teams to both define and implement changes. Zytec, for instance, uses its management by planning (same as **strategic plan**) system to form cross-functional teams. These teams have a strategic quality goal that they are charged with meeting.

In the benchmarked Japanese Deming Prize winners, and several of the Baldrige Award winners, 80 percent of the themes of **quality improvement teams** are passed down from the strategic quality goals. A mere 20 percent of those themes comes from self-initiated efforts to change. Thus, the seeming "bottom-up" system is a bit of a misnomer. Answers come from the bottom. Questions come from the top. The top in this case means what customers have indicated as important needs for improvement.

Bottom-Up Employee Involvement

Baldrige Award examiners like to see a system, such as a **suggestion system,** to help foster bottom-up involvement. The examiners also want to see how results from last year's objectives are rolled into the next year's **strategic planning** process.

Most of the bottom-up efforts are handled through **suggestion systems** and receptive managers. You don't need a whole process or a **quality improvement team** to make a simple improvement. Managers need to be trained to be receptive and to announce the daily accomplishments as well as the major ones. Zytec empowers its people to make a process change with just one other manufacturing person's approval.

Spirit

The intent and spirit with which continuous improvement happens determines its success. Quantum leaps happen one tiny step at a time. Coaching managers and employees to recognize and applaud success helps company cultures move from negative to positive styles. Many companies have special motivational campaigns to encourage this effort. Use of outside consultants at first can coach upper-level managers on how to coach.

Other positive culture setters can also help. Zytec employees pass out beads to each other when they catch each other doing something special. Holiday Hotels does some of the same thing with their positive ambush effort. They acknowledge their winners at an awards dinner. Learning this positive culture is much more important to quality success than all the plans and charts in the world.

Too many companies have a continuous improvement effort that is considered a joke by employees. The prime reason for failure is because managers try to dictate and delegate quality rather than listen to their employees and practice continuous improvement themselves. These same authoritarian managers are usually last to truly empower employees. Improvement teams come up with ideas that never make it past the next layer. This is a killer for continuous improvement.

Suggestion System

One of the easiest start-up exercises is to simply put a box in a prominent place and ask people to drop their suggestions in it. It sounds deceptively simple. It is. In order for a suggestion box to work, it has to have a system behind it. The system entails managers promoting it, a clear process of reviewing the suggestions, a cycle time of the suggestions' receipt to their disposition, and a cultural willingness of managerial change. Celebrations and recognition of contributors are also important.

A **suggestion system** will foster bottom-up improvement and creative thinking. Tricia Kelly reports in *Quality Progress* that the average number of suggestions that an employee makes per year in the United States, industrywide, is 0.1.[48] Toyota maintains that it receives 100–140 ideas per employee per year, of which 97 percent are actioned.[49] When I spoke with Dr. Kano about Toyota, I commented with disbelief about the high implementation rate. I asked, "How can Toyota implement so many? Don't they know many won't work?"

Dr. Kano said, "Yes, but the bias toward implementation fuels creativity and even more suggestions. If a particular suggestion doesn't work out, it generates one that does."[50]

Exhibit 3.26 shows one type of form that can be used in a suggestion system. Design a suitable form for your organization and attach it to the suggestion box or put it on your electronic mail.

Deployment

After you have designed a form for people to follow, set up a review system. Determine who will look at the suggestions, how they will be evaluated, and who will make the final decision about implementation.

Name _____   Department _____
Phone _____   Supervisor _____

Suggestion _____

Problem definition   Drawing:
_____
_____

Estimated cost of problem (include money, morale, opportunity)
_____
_____

Proposed solution
_____
_____

Benefits
_____
_____

Comments
_____
_____

Exhibit 3.26 Sample suggestion form.

1. Flowchart the suggestion review system.
2. Determine the standard for turnaround time from receipt to disposition of the suggestions.
3. Set a bias toward action in the reviewers through thoughtful recruitment of the reviewers and their training.

Set up a recognition or award system. Milliken uses recognition rather than money to reward top suggestions. IBM awards up to $100,000 for beneficial suggestions. Either way works. One caution about money. It can be distracting as people start to focus on the relative worth of suggestions. Other recognition vehicles, such as newspapers, bulletin boards, and meeting announcements frequently work just as well as money.

The criteria for recognition need to be broad. You want to spread out the kudos over as broad a base as possible. In addition to rewarding cost-saving ideas, consider giving awards to such areas as the following:

- Most creative idea(s)
- Most customer-focused idea(s)
- Most technically ingenious idea(s)
- Idea(s) which fostered the most teamwork
- Implementation which took the most effort
- Idea(s) that promote the greatest interdepartmental cooperation

Have top managers promote the system by announcing it in meetings. Gather information on how many suggestions are made and implemented. Keep track and show progress on highly visible charts.

Caveats: Ford Motor Company studied suggestion programs around the world and found that structuring suggestions around cost savings doesn't work. The system must: (1) get employees involved; (2) train employees so they submit good ideas; and then (3) look for cost savings.

Use the review process to mix the various levels of the organization. Rather than just have a top management review committee, put various levels, from management to the frontline people, on the review team. For more complicated ideas, an improvement team may need to be selected to cost and test the solutions. Make sure the idea generator is on the team.

Track your suggestions but don't make the number of the suggestions the focus. One engineering company I worked with set a goal of having 1000 suggestions that year. The person in charge went through all the project summaries done in the past five years and stuffed the suggestion box in order to meet the quota. Word got around to the employees about how suggestions were generated and the **suggestion system** was a joke. Since the manager who set the goal was only interested in numbers, he never knew how the quota was achieved. Only managers who truly care about making a better service or product will produce a viable **suggestion system.**

Quality Improvement Teams

Each organization has its own name for **quality improvement teams.** A **QIT** is a group of people who come together to define quality problems and brainstorm possible solutions. They were known as quality circles in the 1980s. Because the idea behind quality circles was only superficially understood by U.S. managers in the 1980s, there were more failures than successes. **QITs** only work in organizations that have a culture and infrastructure for improvements to be adopted.

Many organizational problems are complex. Reducing defects, increasing cooperation between groups, alleviating communication problems, reducing turn times in inventory, and reducing scrap all may involve several departments. Thus, it takes input from many departments to make an improvement work. Likewise, getting the workers who implement the solution involved in defining potential solutions will circumvent resistance.

Caveats: In Japan, service companies use **QITs** less than manufacturing companies. Improvements in Japanese companies are carried more through a **suggestion system** and **strategic planning** (called *hoshin planning*) than through **QITs.**

Learning how to solve problems within a team takes time. Companies need to train facilitators who can then help groups open up, define problems, and come to viable solutions. As groups learn these facilitation skills, they can become self-directed. Usually the skill evolution follows this pattern.

1. A trained outside facilitator works at the executive level and trains internal facilitators.
2. The trained internal facilitators help team members learn how to facilitate.
3. Team members take turns acting as facilitator.
4. No facilitator is needed because the teams are self-directed.

Don't try to start at the end of this evolution. Facilitating problem-solving meetings is one of the most complex skills in quality. The evolution described here can take months and even years.

How to Form QITs

Usually the **steering committee** defines the **key quality objectives** in the **strategic plan**. The committee then needs to decide what different departments or groups are specifically needed for progress. Those objectives need to be translated to each division or department. Then either a departmental or cross-functional team will be chosen to develop specific goals and an implementation plan.

The **steering committee** might consider the following when choosing members for any **QIT.**

- Appropriate expertise
- Nonconflicting personalities
- Representation from all disciplines involved
- Representation from the people who will be implementing the solution
- Action-oriented people—yet ones who can also pause to define the problem

Many **QITs** have a core group and then call in additional resource people as required.

Logistics

Size. Five to eight members is the ideal number for a **QIT,** but two people may suffice. If more than eight are on a team, decisions may bog down.

Number of meetings. Meeting intervals should be dependent on the urgency of the problem and how reasonable the intervals are for participants. Consider meeting more often at first and then having longer intervals between meetings.

Roles. Both a facilitator and a scribe are necessary. The scribe needs to circulate the agenda and meeting notes.

How to Get a QIT Started

1. **QITs** need to find a management sponsor, or a management sponsor needs to find them.
2. Determine the mission of the group.
3. Have the facilitator and/or **QIT** fill out the QIT project charter. (This outline follows.)

4. Negotiate the QIT project charter with the management sponsor.

Exhibit 3.27 displays a QIT project charter that can be used to plan the work of the **QIT** and get agreement from management. The charter helps the group organize its efforts and specify needs ahead of time.

Date _____ Facilitator _____

Mission of the QIT _____
_____
_____

Who initiated this project?
_____

Benefits of completing this mission (and/or what goals it relates to in the **strategic plan**).
_____
_____

People or functions who need to be involved

In problem definition	In implementation	From management
1. _____	1. _____	1. _____
2. _____	2. _____	2. _____
3. _____	3. _____	3. _____
4. _____	4. _____	4. _____
5. _____	5. _____	5. _____

Project selection checklist (necessary elements in a successful project)

❐ Project relates to **key quality objectives**   ❐ Implementors are not being reorganized

❐ Project has impact on external customers   ❐ Project has visibility in company

❐ Project has a high-level sponsor   ❐ Project has a starting and ending point

❐ Project is only being studied by this QIT   ❐ Project is a process, not a solution

❐ Process is within the control of the group   ❐ Project is not too complex

❐ Return on investment is worth the time   ❐ Project focuses on one or two processes only

---

Exhibit 3.27 QIT project charter.

QIT Project Methodology

Exhibit 3.28 outlines a sequence of steps typically taken as a QIT moves through scoping and resolving problems. The six-step project methodology allows the group to also think through the tools that

General steps	Specific steps	Tools*	Who	Due
1. Identify and scope the problem. (Scope means to assess when, where, and how it happens.)	1. 2. 3. 4. 5.	1. 2. 3. 4. 5.	1. 2. 3. 4. 5.	
2. Analyze causes.	1. 2. 3. 4.	1. 2. 3. 4.	1. 2. 3. 4.	
3. Brainstorm solutions.	1. 2. 3.	1. 2. 3.	1. 2. 3.	
4. Take corrective action.	1. 2. 3.	1. 2. 3.	1. 2. 3.	
5. Evaluate effect.	1. 2. 3.	1. 2. 3.	1. 2. 3.	
6. Standardize.	1. 2. 3.	1. 2. 3.	1. 2. 3.	

*Types of tools

Customer satisfaction	*Team process/analysis*	*SPC*
Focus group	Brainstorm	**Control charts**
Survey	Nominal group technique	**Check sheets**
Feedback tool	**Fishbone diagram**	**Pareto charts**
Observation	Flowchart	Scatter diagrams
Hard measure (turnover)		

Exhibit 3.28 Six-step project methodology.

may be relevant, who will do it, and when it is due. Exhibit 3.29 can be used to format your own key steps and time line.

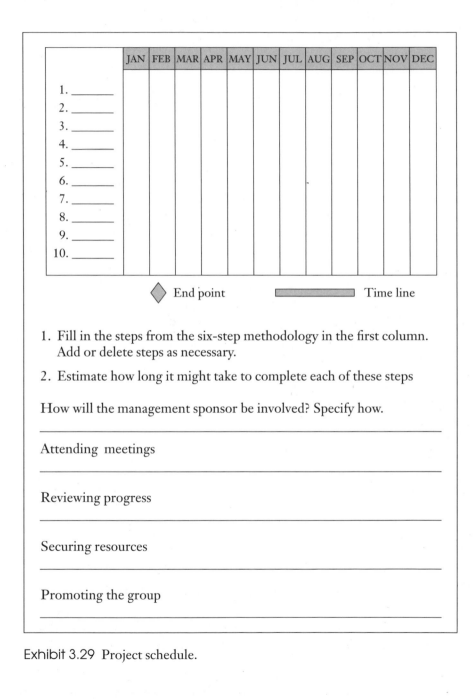

	JAN	FEB	MAR	APR	MAY	JUN	JUL	AUG	SEP	OCT	NOV	DEC
1. _____												
2. _____												
3. _____												
4. _____												
5. _____												
6. _____												
7. _____												
8. _____												
9. _____												
10. _____												

◆ End point   ▬▬▬▬▬ Time line

1. Fill in the steps from the six-step methodology in the first column. Add or delete steps as necessary.

2. Estimate how long it might take to complete each of these steps

How will the management sponsor be involved? Specify how.

_____

Attending meetings

_____

Reviewing progress

_____

Securing resources

_____

Promoting the group

_____

Exhibit 3.29 Project schedule.

Employee Education and Training Plan

Employee education and training needs to go beyond the basics of their job skills. The training is yet another means of enhancing those processes that are critical to customer satisfaction. At Prism Radio Partners, a broadcasting company, we found customers wanted salespeople to help them with marketing ideas and plans. Thus, we developed a seminar and skill entitled "Integrated Marketing Specialist." The training plan needs to start by looking at the strategic quality objectives and figure out what training is needed (see Exhibit 3.30). Exhibit 3.31 lists training topics that are typical for different organizational levels.

Who (individual/ department)	What training	KQO	When	Budget	Comments
1.					
2.					
3.					
4.					
5.					
6.					
7.					
8.					
9.					
10.					

Write the number of the key quality objective that relates to the training.

Key quality objective
1. _____
2. _____
3. _____

Exhibit 3.30 Training plan.

Executives (40 hours)	Middle management (40 hours)	Frontline (20–40 hours)
Quality overview	Quality overview	Job skill seminars
Quality culture	Quality culture and concepts	Language (if needed)
Quality leadership	Statistical process control	Math (if needed)
Strategic plan	Team building	Quality overview
Statistical process control	Quality systems	Statistical process control
Quality systems	**Customer satisfaction**	Team problem solving
Customer satisfaction	**Quality function deployment**	
Quality function deployment		

Exhibit 3.31 Sample quality training topics.

Usually the executives and/or quality **steering committee** are trained by an outside facilitator or at outside seminars. The outside facilitators should then train the trainers within the organization.

Remember, basic skills come first. If employees need help just doing their jobs, speaking English, or doing basic math, you need to work on those skills first.

Employee Performance and Recognition*

The HR department typically controls many of the performance review and recognition systems. These areas were covered under the **quality accountability system,** since leadership must drive the overall programs for them to work. HR usually ends up implementing forms, meetings, bulletin boards, and the like. The overall plan, design of performance, and recognition must be integrated though with the **strategic plan.** Otherwise, all will fail.

Survival in the 1990s depends on employee innovation. The Manufacturer's Alliance for Productivity and Innovation (MAPI) found that companies that empower employees are twice as likely as other firms to report significant product or service improvement. Fifty-two percent of the companies reporting well-integrated employee involvement and quality programs cited significant improvements in the overall quality of their products or services. Of those companies that did not have well-integrated programs, only 16 percent reported

*Also see **Quality Accountability System**

similar quality gains. Companies with established programs were also more likely to report better results in profitability, internal costs, errors, and employee morale.[51] The more fun and imaginative you can make recognition of innovations, the more creative you will encourage your employees to be.

Measures of Employee Well-Being and Satisfaction

Baldrige Award examiners want to see that managers take a sincere interest in developing their employees and looking after their well-being. Employee satisfaction measures and goals are an important part of this. If you are going for the gold medal, the Baldrige Award examiners will look for results and extras. As the following indicate, perception of satisfaction is not enough. Examiners want to see hard measures of employee well-being. As with all the other categories, managers need to set goals, monitor progress, and show results. Types of well-being and satisfaction issues include the following:

- Health
- Safety
- Aesthetics
- Ergonomics
- Counseling
- Job rotation
- Housekeeping
- Day care centers
- Grievance procedures
- Recreational facilities

Employee Satisfaction and Improvement Goals and Measures

How do you determine what questions to include in employee satisfaction measures? Do you have a systematic plan that will help your organization move forward on specific measures? Are you actually using the data that you collect to make changes? What are the results over time of those changes? Consider both hard and soft measures of employee well-being.

Hard measures include data on strikes, safety, turnover, evidence of cultural diversity, absenteeism, and accidents. Industry standards should be added to any measure you make to show how you compare to others. Soft measures include surveys, advisory groups, **focus groups**, and so on.

Baldrige Award examiners will want results to be substantiated—preferably by graphs that show progress over time. Exhibit 3.32 provides an example of data condensed in graphic form.

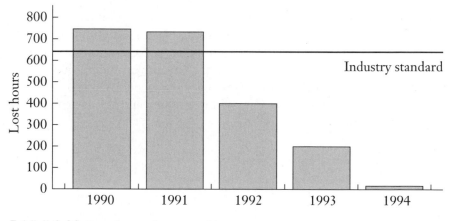

Exhibit 3.32 Lost hours due to accidents.

6.0

Process Management

Process management considers what the organization does to continuously improve its service or product quality as well as operations that lead to that quality. Key areas to consider include the following:

- Research and development
- Design of new products and services
- Management of supplier quality
- Improvement of products and services
- Betterment of the improvement process
- Management of the quality of support services, such as purchasing, sales, accounting, legal, finance, and so on.

New (or Improved) Product (or Service) Design System

A **new product design system** is a clear set of steps used to: assess client requirements; see that requirements are addressed in design; have multidisciplinary involvement early in design; see that requirements have been integrated; and test prototypes before production. Design can also include facilities to build products or deliver services.

Having a well-planned design system can curtail a myriad of headaches and disappointments. Downstream costs include returned goods and lost customers. These costs are averted when more attention is paid to careful customer needs analysis, management of design

119

changes, and involvement of all the affected departments in design. (This is called concurrent engineering.) In addition, premature release before errors have been discovered or corrected can create a poor quality reputation. For example, you have probably seen several computer and software companies promise great new technological advances and release products prematurely to disappointed customers. Even some airlines have promised new services like priority baggage handling before their systems were in place to support it. Thus, expectations are raised and customers are needlessly disappointed.

The following represents the basic ingredients necessary in a design system.

1. How customer requirements are assessed

2. How customer requirements are translated into specifications

3. How national standards or government regulations are incorporated

4. How design validation ensures these requirements are met

5. How the change process is controlled and documented

6. How prototypes are developed

7. How prototypes are field tested before production

8. How processes 1–7 are continuously improved

9. How early design reviews are done by multiple disciplines (sales, design, environmental, finance, suppliers, maintenance, operations, waste reduction, production, and so on) in order to optimize performance criteria, process capability, supplier capabilities, customer responsiveness, and ecology of design

Tools

Quality function deployment is one of the primary tools used for **new product designs.** A second tool is design review and control. A major engineering company reported at the Orlando Energy Quality conference in September 1992 that most of its drafting errors could be traced to cracks in the change process. By clearing up the change process, the engineering company improved its defect rate by over tenfold. Design review and control is too specific to each industry to be detailed here. The main ingredients of the process are the identification of who, what, when, and how.

Quality Function Deployment

Quality function deployment (QFD) is a tool to translate client requirements into specifications for new or improved products or services. It is also known as the house of quality. It incorporates how the competition ranks on each dimension. This focus on clients and competition at the same time helps designers "go the extra mile."

Benefits of QFD include the following:

- QFD can be used from educational settings to high-tech service companies. Starting with the needs of customers assists the company to maintain client focus.

- QFD forces sales to learn more about engineering and operations and vice versa. These team decisions are more efficient and effective in design.

- QFD offers a convenient way of organizing complex information so decisions are based on data, not just gut-level feelings. The following simplified house of quality can be used to translate client requirements into service/product specifications.

Exhibit 3.33 provides a simplified way of listing client needs; assigning weights to the importance of those needs; detailing the specifications that relate to those needs; and comparing your strengths and weaknesses on those dimensions to your competitors.

Production and Delivery Quality Management Processes

Back in the section on **strategic planning**, the **key processes** and **key quality indicators** were discussed. This section refers back to those **key processes** and **key quality indicators** and tells how to maintain and improve their performance. If your company is a hospital, **key processes** might be emergency transportation services, admissions, nursing, food, insurance, and so on. The **key quality indicators** specify which of these are critical to your success. **Key quality objectives** are improvement goals for these areas. This section merely reiterates what you are doing to monitor progress on each of these targets.

Client needs	Weight	Spec #1	Spec #2	Spec #3	Competitors A B C
1.					
2.					
3.					
4.					
5.					
6.					

Client needs: Anything related to ease of use, delivery times, ordering
                system, billing system, safety, reliability, reporting systems,
                information systems, technical expertise, maintenance, and
                so on.
Weight: Strength of need. Scale is 1 to 10; 1 is low and 10 is high.
Specs: Translates into service design. A delivery time may be
                detailed as within two hours.
Competitors: Clients' ratings of competitors on this same dimension. Scale
                is 1 to 10; 1 is low and 10 is high.

Exhibit 3.33 Simplified quality function deployment.

Key Ingredients

Every company is different in how it monitors processes to make sure
that its service or product is performing according to specifications.
Key ingredients of this vigilance and corrective action include controlling processes, dealing with out-of-control processes, incorporating
the fixes, and further improvements.

Controlling Processes. This includes determining the capability of
each critical service or manufacturing process through tools such as
control charts, check sheets, and machine vision (computerized
comparison of ideal digital image to the widget being monitored on
the conveyer belt). In describing these control mechanisms specify the
key processes, which are key indicators of quality and operational

performance and requirements; and the ways quality and operational performance are determined, including in-process and end-process measurements.

Dealing with Out-of-Control Processes. What happens when a process is not meeting specifications? How are root causes determined so what caused the defect is changed? Usually companies talk about assigning a **quality improvement team** and using a six-step methodology to fix the problem.

Incorporating the Fixes. Sometimes problems are studied but solutions are positive but never incorporated into the operation. Baldrige Award examiners will look for such areas as resetting process limits or standards, verifying improvements, and deploying improvements throughout the company.

Further Improvements. The monitoring and correction process also needs continuous improvement. How does this happen? Baldrige Award examiners will look for evidence of areas such as the following:

- Waste reduction
- Process simplification
- **Benchmark** information
- Process research and testing
- Testing of alternative technologies

Support Services Quality System

Many manufacturing organizations have incorporated total quality management on the factory floor but have neglected the services that support manufacturing. Yet a 1986 study of consumer behavior at the Technical Assistance Research Program Institute of Washington D.C. found that 75 percent of customers that defect from products do so because of the services attendant to that product. Examples include the following:

- Delivery is slow.
- Sales service is poor.

- Invoices are inaccurate.
- Engineering modifications are unresponsive.
- Service of the product after it breaks is irritatingly slow or unresponsive.

Thus, a high-quality company applies the same rigor of continuous improvement to these support services as it does to its primary production. Typical support services include the following:

- R & D
- Sales
- Accounting
- Purchasing
- Public relations
- Software services
- Finance

- Personnel
- Marketing
- Plant facilities
- Information services
- Secretarial or administrative services

Design of a Support Services Quality System

Each support group would go through the following sequence.

1. Receive overview training in total quality manufacturing or service.
2. Select **steering committee** and **champion(s).**
3. Do **customer needs assessment** and satisfaction baseline.
4. Determine **key processes** and **key quality objectives**
5. Roll 1–4 into the **strategic plan.**
6. Set up appropriate QITs to do problem definition and resolution. At first, use trained facilitators to help these teams.
7. Set up monitoring systems, correct target problems, and celebrate.
8. Feed results back into the annual **strategic planning** process.

Supplier Quality Management System

Suppliers assume varying importance in the overall quality level of an organization. If the supplies consist of office materials, such as pencils, pens, and paper, this section may not be vital. If the suppliers provide basic ingredients to the product or service quality, such as airplane mechanics, physicians, contract faculty, and so on, then managing supplier quality is critical.

The following are centerpieces to this system:

1. Communication of **key quality indicators** and requirements to suppliers
2. Methods of monitoring and correcting deviations from these requirements. Examples might be the following:
   a. Supplier audits
   b. Inspection at the supplier's location or the receiving dock
   c. Certification of suppliers that meet certain standards
   d. Testing and rating systems
3. How the organization improves its own purchasing activities. Included in here might be the following:
   a. How supplier performance is tracked
   b. How suppliers are selected based on their previous performance
4. How you help suppliers improve their quality. Choices include the following:
   a. Consulting suppliers
   b. Inviting them to your quality training programs
   c. Joint **strategic planning** sessions
   d. Partnerships (limiting suppliers to just the best and having long-term relationships with them)
   e. Incentives and recognition (supplier award ceremonies, certificates, and so on)

Quality Assessment System

One of the difficulties most organizations have is honesty. Human nature is to report results to managers in a positive nature to make the department or division look good. Having neutral people come in to do quality audits helps preserve the honesty of the organization. Then doing honest self-assessments shows that continuous improvement has been applauded and internalized, instead of done as a token activity.

The approach to quality assessment reiterates the implementation plan used in the **strategic plan**. What is used to examine the organization's systems, processes, and practices? What is used to examine the end product or service? This section merely summarizes all the assessment devices and how they get rolled into the **strategic planning** process. Usually it is a multilevel approach that includes many of the following examples.

- Strategic quality review process
- Customer needs assessments
- Customer satisfaction databases (complaint tracking, correction tracking)
- Internal quality audits or self-assessments
- Audits from independent organizations
- Mystery shopper programs
- Lost customer surveys

Each assessment device should indicate how often it is done and who does it. Exhibit 3.34 provides an example.

Use of Data. This section talks about how the results of the customer satisfaction system get turned into goals, improvements, and fixes. The **strategic planning** process should be the pivot point in this discussion.

Documentation. The trick to documentation is to keep it simple. Records can help deploy the improvements in marketing and the rest of the organization. Tracking devices make sure you can keep tabs on the complexity of all of the pieces.

Assessment devices	How often	Who performs
Customer surveys	Yearly	Quality staff
Mystery shoppers	Monthly	Mystery shopper services
Complaint tracking	Daily	Customer service

Exhibit 3.34 Customer satisfaction system.

Success Stories. What records are kept along the way? Each **quality improvement team** should keep a QIT project charter and simple (one page is plenty) history of its process. An illustration was shown in the logical sequence of tools. Only the solution and evidence of a fix need to be added. These become the **Success Stories.**

Tracking Devices. Computerized or manual tracking systems are helpful for customer complaints, corrective actions, and customer defections.

7.0

Business Results

Increasing emphasis has been placed in the Baldrige Award criteria on business results. Are all of your efforts paying off? Is your organization improving?

Results you report should be relevant to your **strategic plan.** Just reporting any results makes it seem as if you either forgot about your goals or had something to hide. In addition to relevance, Baldrige Award examiners look for three different strengths in the results you report.

1. Are the numbers high (or low, if defects)?

2. Is the variation low in your numbers?

3. Is the trend one of continuous improvement?

High numbers are measured both in real terms and against competitors. If your customer satisfaction scores are 65 percent favorable, that seems low. If your competitors, though, rank a high of 55 percent, your score gets boosted upward. If you have reported data from the last 10 years and your numbers jump all over, variation is considered high.

The trend indicates your progress over time. A trend of improvement sustained over five years is considered noteworthy.[52]

Business results need to be reported in the following areas.

• Product/service quality

• Company operational

- Business process and support service
- Supplier quality

Exhibit 3.35 is an example of a succinct way to show a snapshot of the organization. This will help you show off your results to customers, employees, and shareholders. Remember, graphs of results help increase the usefulness of your data.

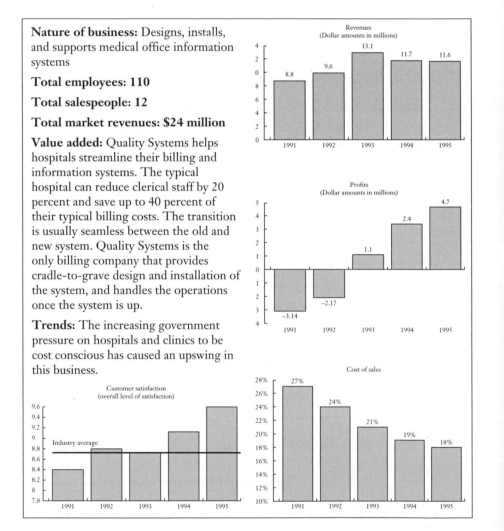

Exhibit 3.35 Quality Systems corporate snapshot.

Summary

The first section of this book outlined the steps of what to do in a world-class quality organization. The second part provided tips on how to do it as seen in many high-performing companies. Both sections followed the Baldrige Award criteria in logic and requirements so that companies could readily make application if they chose to at a later point. Small companies may want to eliminate many of the steps if they are unnecessary.

Many companies do very well at several of the ingredients that go in to a high-performing company, but they are weak in others. The intent of this book is to simplify a complicated effort as much as possible so that managers could get an idea of the breadth and depth of involvement required at all levels.

Ideally, U.S. colleges would change their MBAs to be structured around this proven template for excellence. Traditional focus on finance would be coupled with a competitive focus on quality. Both are vital to survival in a highly competitive global economy that is changing the way managers think and work.

This book has focused on implementation practices of successful companies. The personal style and enthusiasm of the internal leaders are equally as important. The side benefit of putting these concepts and systems in place is that people finally fix the problems they have been living with for years. The fixes save wasted time in redundant steps, unnecessary meetings, getting approval signatures, filling out redundant reports, and seeking inaccessible information. That time is freed up to truly understand what the customers want, to be creative about designing new services and products, and in implementing the plan with spirit. This extra measure of spirit and service is so noticeable to customers that the service or product becomes irresistible. Being irresistible is my ultimate wish for all of you.

Notes

1. Noriaki Kano, interview by author, Tokyo, Japan, 28 September 1992.

2. Jon Brecka, "Employee Involvement Linked to Quality Gains," *Quality Progress* (December 1993): 14.

3. Jon Brecka, "Corporate Leaders Are Not Ready for Changes," *Quality Progress* (December 1993): 12.

4. W. Edwards Deming, "Plan for Action for the Optimization of Service Organizations," (paper presented at Dr. Deming's Plan for Action for the Optimization of Service Organizations conference, San Jose, Calif., 21–22 July 1992).

5. Nancy Mann, interview by author, San Jose, Calif., 21–22 July 1992.

6. Hirobu San, head of the Komatsu Career Council, interview by author, Tokyo, Japan, 26 September 1992.

7. David Kearns and David Nadler. *Prophets in the Dark: How Xerox Reinvented Itself and Beat Back the Japanese* (New York: Harper Business, 1992), 201.

8. Chuck Roberts, vice president of quality at Ames Rubber, interview by author, Washington, D.C., 6 February 1994.

9. Kearns and Nadler, *Prophets in the Dark*, 237.

10. Brooks Carder and James Clark, "The Theory and Practice of Employee Recognition," *Quality Progress* (December 1992): 25–30.

11. Jeremy Main, *Quality Wars: The Triumphs and Defeats of American Business* (New York: Free Press, 1994), 250.

12. W. Edwards Deming, speech at Quality, Productivity, and Competitive Position conference, Pasadena Calif., 17–20 November 1992).

13. Doug Tersteeg, Zytec quality manager, interview by author, Redwood Falls, Minn., 22 October 1992.

14. Judy Stewart, president of Rancho Vista Bank, interview by author, Vista, Calif., 14 February 1994.

15. Anisia L. Kowalchuck, AT&T High Performance Quality Consultant, interview by author, Washington, D.C., 5 February 1995.

16. Nacchio, interview.

17. Main, *Quality Wars*, 84.

18. Jo Sanders, customer service manager at Wainwright, Washington, D.C., 7 February 1995.

19. Karen Bemowski, "To Go Boldly Where so Many Have Gone Before," *Quality Progress* (February 1995): 29–34.

20. Main, *Quality Wars*, 83.

21. Michael Maslak, president of North Island Federal Credit Union, interview by author, Pasadena, Calif., 16 November 1994.

22. Kano, interview.

23. Marilyn Carlson, vice president of sales, GTE Directories, interview by author, Washington, D.C., 7 February 1995.

24. Mark Graham Brown, *Baldrige Award–Winning Quality* (Milwaukee: ASQC Quality Press, 1994), 284–297.

25. Earnest W. Deavenport, CEO of Eastman Chemical, interview by author, Washington, D.C., 7 February 1994.

26. Bob E. Hayes, *Measuring Customer Satisfaction* (Milwaukee: ASQC Quality Press, 1992), 55. Also see Alan Dutka, *AMA Handbook for Customer Satisfaction* (Lincolnwood, Ill.: NTC Business Books, 1994), p. 91, where he says, "Therefore a 7–10 point scale will not discriminate any better than a 5 point scale."

27. Bradley T. Gale, *Managing Customer Value* (New York: Free Press, 1994), 208.

28. Pamela Gordon, "Customer Satisfaction Research Reaps Rewards, *Quality* (May 1993): 39–41.

29. Larry Armstrong and William C. Symonds, "Beyond 'May I Help You,'" *Business Week*, 25 October 1991, 102.

30. Terence Pare, "How to Find Out What They Want," *Fortune* (autumn/winter 1992): 39.

31. K. Shimo Yamada, Spa Resort Hawaiian, interview by author, Tokyo, Japan, 5 October 1992.

32. Sanders, interview.

33. Jonathan Weber, "Computer Services for Hire: Outsourcing Thrives as Firms Shed Data Processing Chores," *Los Angeles Times*, 2 August 1992.

34. Kearns and Nadler, *Prophets in the Dark*, 238.

35. Bemowski, "To Boldly Go," 29–34.

36. Main, *Quality Wars*, 120.

37. Tersteeg, interview.

38. Main, *Quality Wars*, 121.

39. Michael J. Spendolini, *The Benchmarking Book* (New York: AMACOM, 1993).

40. W. Edwards Deming, *Out of the Crisis* (Cambridge, Mass.: MIT Center for Advanced Engineering Study, 1986), 55.

41. Jacqueline Graves, "Management Tools That Work," *Fortune*, 30 May 1994, 15.

42. Michael Traecy, *Discipline of Market Leaders* (Reading, Mass.: Addison-Wesley, 1995), 1–14.

43. Jennifer Laabs, "Ben & Jerry's Caring Capitalism," *Personnel Journal* (November 1992): 50–57.

44. Earnest W. Deavenport, interview at the Quest for Excellence Conference, Washington, D.C., 7 February 1994).

45. Kano, interview.

46. "Companies Have Increased Employee Training Budgets," *Quality Progress* (December 1991): 12.

47. Hirobu San, interview.

48. Tricia Kelly, "Keep On Keeping On," *Quality Progress* (April 1993): 20.

49. Philip Atkinson, *Creating Culture Change: The Key to Successful Total Quality Management* (San Diego: Pfeiffer & Company, 1990), 243.

50. Kano, interview.

51. Jon Brecka, "Employee Involvement Linked to Quality Gains," *Quality Progress* (December 1993): 14.

52. Brown, 284–297.

Part 4

Resources

Bibliography and Books
That Highlight
Service Quality

Aguayo, Rafael. *Dr. Deming: The American Who Taught the Japanese About Quality*. New York: Carol Publishing Group, 1990.

Anderson, Kristin, and Ron Zemke. *Delivering Knock-Your-Socks-Off Service*. New York: American Management Association, 1991.

Armstrong, David. *Management by Storying Around: A New Method of Leadership*. New York: Doubleday, 1992.

Armstrong, Larry, and William C. Symonds. "Beyond 'May I Help You.'" *Business Week*, 25 October 1991, 100–103.

Atkinson, Philip. *Creating Culture Change: The Key to Successful Total Quality Management*. San Diego: Pfeiffer and Company, 1990.

Barker, Joel Arthur. *Future Edge: Discover the New Paradigms of Success*. New York: William Morrow and Company, 1992. (Centerpieces are anticipation, innovation, and excellence. Upper-level management needs to read.)

Brassard, Michael. *The Memory Jogger Plus: Featuring the Seven Management and Planning Tools*. Methuen, Mass.: GOAL/QPC, 1989. (Helps with application of principles.)

Brown, Mark Graham. *Baldrige Award Winning Quality*. Milwaukee: ASQC Quality Press, 1995.

Byrne, Bill. *Habits of Wealth: Delight Employees and You Will Delight Customers*. Sioux Falls, S. Dak.: Performance One Publishing, 1992.

Camp, Robert. *Benchmarking: The Search for Industry Best Practices That Lead to Superior Performance*. Milwaukee: ASQC Quality Press, 1989.

Cannie, Joan Koob. *Keeping Customers for Life*. New York: American Management Association, 1991. (Cannie goes into the economics of customer dissatisfaction, gives examples of client surveys, and discusses a service business plan.)

Clemmer, Jim, and Barry Sheehy. *Firing on all Cylinders: The Service-Quality System for High-Powered Corporate Performance.* Homewood, Ill.: Business One Irwin, 1992. (This book is one of the best on total quality service. Although it is not easy reading, the book provides more research to validate claims than most books. It should be a primer for all those who are embarking on this mission. Upper-level management needs to read.)

Collins, James C., and Jerry I. Porras. *Built to Last: Successful Habits of Visionary Companies.* New York: Harper Collins, 1994.

Condon, John C. *With Respect to the Japanese.* Yarmouth, Maine: Intercultural Press, 1984.

D'Egidio, Franco. *The Service Era: Leadership in a Global Environment.* Cambridge, Mass.: Productivity Press, 1990.

Deming, W. Edwards. *Out of the Crisis.* Cambridge, Mass.: MIT Center for Advanced Engineering Study, 1986.

Denton, Keith. *Quality Service.* Houston: Gulf Publishing, 1989.

Denton, Keith, and Barry Wisdom. "The Learning Organization Involves the Entire Workforce." *Quality Progress* (December 1991): 69-72.

Desatnick, Robert. *Keep the Customer! Making Customer Service Your Competitive Edge.* Boston: Houghton Mifflin, 1987. (This book covers strategies used by McDonald's and Citicorp to achieve service performance. It focuses on hiring and setting service standards.)

Drucker, Peter F. *Managing in Turbulent Times.* rev. ed. New York: Harper Business, and Newman, Mass.: Butterworth-Heinemann, 1993.

Dutka, Alan. *AMA Handbook for Customer Satisfaction: A Complete Guide to Research Planning and Implementation.* Lincolnwood, Ill: NTC Business Books, 1994.

Dwyer, Paula, and Dave Griffiths. "How the Pentagon Can Go to War Against Abuse." *Business Week,* 4 July 1988, 34.

Ealy, Lance A. *Quality by Design: Taguchi Methods and U.S. Industry.* Dearborn, Mich.: ASI Press, 1988.

Ernst and Young. *International Quality Study: The Definitive Study of the Best International Quality Management Practices.* Cleveland, Ohio: Ernst and Young, 1993.

Futami, Ryoji. *Graphical Methods for TQC.* Tokyo: JUSE Press, 1985.

Gabor, Andrea. *The Man Who Discovered Quality: How W. Edwards Deming Brought the Quality Revolution to America—The Stories of*

Ford, Xerox, and GM. New York: Times Books Random House, 1990. (Upper-level management needs to read.)

Gale, Bradley T. *Managing Customer Value.* New York: Free Press, 1994.

Goodman, John, et al. "Ineffective—That's the Problem with Customer Satisfaction Surveys." *Quality Progress* (May 1992):39–41.

Gordon, Pamela. "Customer Satisfaction Research Reaps Rewards." *Quality* (May 1993): 39–41.

Graves, Jacqueline. "Management Tools That Work." *Fortune,* 30 May 1994, 15.

Hayes, Bob E. *Measuring Customer Satisfaction: Development and Use of Questionnaires.* Milwaukee: ASQC Quality Press, 1992. (Helps with application of principles.)

Imai, Masaaki. *Kaizen: The Key to Japan's Competitive Success.* New York: Random House Business Division, 1986.

Ishikawa, Kaoru. "The Quality Control Audit." *Quality Progress* (January 1987): 39–41.

————. *What is Total Quality Control? The Japanese Way.* Translated by David J. Lu. Englewood Cliffs, N.J.: Prentice Hall, 1985.

Japanese Human Relations Association Staff. *The Service Industry Idea Handbook: Employee Involvement in Retail and Office Improvement.* Cambridge, Mass.: Productivity Press, 1990.

Japanese Standards Association Staff. *An Easy Approach to the Seven Tools of QC.* White Plains, N.Y.: Quality Resources, 1980.

Japan Management Associates Staff. *Kanban, Just in Time at Toyota: Management Begins at the Workplace.* Translated by David J. Lu. Cambridge, Mass.: Productivity Press, 1989.

Juran, J. M. *Juran on Leadership for Quality. An Executive Handbook.* New York: Free Press, 1989.

————. *Just in Time: A Practical Approach on Leadership for Quality.* New York: Free Press, 1989.

Juran, J. M., and Frank M. Gryna. *Juran's Quality Control Handbook.* 4th ed. New York: McGraw-Hill, 1988. (Helps with application of principles.)

Kearns, David, and David Nadler. *Prophets in the Dark: How Xerox Reinvented Itself and Beat Back the Japanese.* New York: Harper Business, 1992.

Keller, Maryann. *Rude Awakening: The Rise, Fall, and Struggle for Recovery of General Motors.* New York: Harper Business, 1990.

Kessler, Sheila. *Measuring and Managing Customer Satisfaction: Going for the Gold.* Fountain Valley, Calif.: Competitive Edge, 1994. (Upper-level management needs to read.)

————. *Team Selling*. Fountain Valley, Calif.: Competitive Edge, 1992. (Upper-level management needs to read.)

Kobayashi, Koji. "Quality Management at NEC Corp." *Quality Progress* (April 1986): 18–23.

Laabs, Jennifer. "Ben & Jerry's Caring Capitalism." *Personnel Journal* (November 1992): 50–57.

Lafferty, John. "Customer Surveys Don't Ask the Right Questions." *Quality Progress* (September 1992): 8.

LeBoeuf, Michael. *How to Win Customers and Keep Them for Life*. New York: Berkeley, 1989. (Lively, simple, and direct book on combining service and sales. Cally Curtis has a two-part video based on this book.)

Liswood, Laura. *Serving Them Right: Innovative and Powerful Customer Retention Strategies*. New York: Harper Business, 1990. (Liswood gives some practical tips on how to keep customers. The tips vary from handling customer complaints to being more strategic.)

Main, Jeremy. *Quality Wars: The Triumphs and Defeats of American Business*. New York: Free Press, 1994.

Mann, Nancy. *The Keys to Excellence: The Story of the Deming Philosophy*. Los Angeles: Prestwick Books, 1985.

Ozeki, Kazuo. *Handbook of Quality Tools: The Japanese Approach*. Cambridge, Mass.: Productivity Press, 1990.

Quick, Thomas L. *Successful Team Building*. New York: AMACOM, 1992.

Rosenbluth, Hal F., and Diane McFerrin Peters. *The Customer Comes Second and Other Secrets of Exceptional Service*. New York: William Morrow and Company, 1994.

Scherkenbach, William W. *The Deming Route to Quality and Productivity. Road Maps and Roadblocks*. Rockville, Md.: CEE Press Books, 1986.

Scholtes, Peter. *The Team Handbook: How to Use Teams to Improve Quality*. Madison, Wisc.: Joiner Associates, 1988. (*The Team Companion* is an instructional kit, including transparencies, reader guides, posters, and so on. Helps with application of principles.)

Schonberger, Richard. *Japanese Manufacturing Techniques: Nine Hidden Lessons in Simplicity*. New York: Free Press, 1982.

Sullivan, L. P. "Quality Function Deployment." *Quality Progress* (June 1986): 39–50.

Tague, Nancy. *Quality Toolbox*. Milwaukee: ASQC Quality Press, 1995.

Timm, Paul R. *50 Powerful Ideas You Can Use to Keep Your Customers.* Harthorne, N.J.: The Career Press, 1992.

Torki, Neviene. *The Link: Statistical Techniques-Process Improvement.* Victoria, Australia: Imageset, 1992.

————. *Construction and Interpretation of Control Charts for Quality Management.* Victoria, Australia: Imageset, 1990.

Treacy, Michael, and Fred Wiersema. *The Discipline of Market Leaders: Choose Your Customers, Narrow Your Focus, Dominate Your Markets.* Reading, Mass.: Addison-Wesley, 1995.

Tribus, Myron. *Deployment Flowcharting: 1989.* Los Angeles: Quality and Productivity, 1989.

Wadsworth, Harrison M., et al. *Modern Methods for Quality Control and Improvement.* New York: John Wiley & Sons, 1986.

Walton, Mary. *The Deming Management Method.* New York: Dodd Mead, 1986.

Weber, Jonathan. "Computer Services for Hire: Outsourcing Thrives as Firms Shed Data Processing Chores." *Los Angeles Times,* 2 August 1992.

Winn, Sharon. *Strategy Definition and Deployment: Align. Enable. Empower.* Bainbridge Island, Wash.: Winn & Associates, 1995.

Zemke, Ron, and Chip Bell. *Service Wisdom: Creating and Maintaining the Customer Service Edge.* Minneapolis: Lakewood Books, 1989.

Zemke, Ron, and Dick Schaaf. *The Service Edge: 101 Companies That Profit from Customer Care.* New York: Penguin Books, 1995.

Glossary

Many of the terms in this glossary have multiple meanings. The definitions provided reflect the way the words are used in this book.

360 degree review Performance review that includes feedback from superiors, peers, subordinates, and clients (internal or external).

Annual quality report The annual quality report focuses on quality performance. The traditional annual reports focus on financial performance. They can be separate or integrated documents.

Approach A Baldrige Award criteria term for the way quality is designed into a process.

ASQC American Society for Quality Control; one of the leading organizations for quality control, quality assurance, and total quality management. Publishes *Quality Progress* monthly.

Baldrige Award A highly competitive, prestigious yearly award given to a maximum of two service, two manufacturing, and two small businesses that have demonstrated excellence in meeting the requirements of the rigorous Baldrige Award criteria. The U.S. Department of Commerce's National Institute of Standards and Technology is the award's sponsoring organization.

Benchmarking The study of processes or systems in another company or unit.

Bottom-up improvement Improvement that emanates from suggestions from the frontline or low-level people in the organization.

Brainstorming A nonjudgmental process for obtaining ideas from a multiparticipant meeting.

Celebration system   A system that budgets, assigns responsibilities, and sets goals for celebrating successes in the organization.

Champions   A manager who oversees specific quality improvement projects and aids them in obtaining appropriate resources and buy-in. Same as sponsor.

Check sheets   A quality tool that has a list of items on the left that are tallied on the right. It is typically used to scope the size and extent of a quality problem.

Checklist   An aid to the memory that itemizes necessary steps or ingredients.

Cluster analysis   A tool used to determine themes within qualitative information, like comments on a customer survey.

Competitor analysis   Analysis of the key competitor's services, products, processes, and prices. Since customers evaluate services against competitors' offerings, each company needs to do likewise.

Complaint tracking   Detailing when complaints come in, what is done about them, and when they are closed. Several software systems exist to aid complaint tracking.

Control chart   W. A. Shewhart's chart for a continuing test of statistical significance. The chart reflects what variation is common cause and what is special cause so that managers know when to take corrective action.

Cost of quality   Used in many ways in quality literature. Loosely interpreted to mean the costs that could be saved if the quality was perfect, no waste existed in the system, and cycle time was at a minimum.

CPI   Continuous process improvement.

CQI   Continuous quality improvement.

Critical processes   High-risk processes. The risk could be to human life or safety or financial.

Cross-functional team   A quality improvement team that consists of representatives from different departments and/or layers of the organization. Many functions (like order fulfillment) cross departmental lines, and need to involve various functions to analyze problems and achieve goals.

Culture   Values that permeate the organization. A culture is communicated by hero stories, by the reasons people get promotions and recognition, by hall talk, and by the questions that are asked by upper management. A total quality service culture is one that is rigorous and customer-focused and that values employees.

Customer focus   Focus on what the customers need and prefer. At high-performing companies, this focus extends to "delighting" customers.

Customer satisfaction   Meeting or exceeding the needs of customers.

Customer satisfaction surveys   Surveys done in writing or by phone to measure the satisfaction levels of either internal or external customers.

Customer value   A term coined by Bradley Gale in *Managing Customer Value* that relates the customer's perception of the services relative to the competition and price paid for those services.

Data driven   Using data to make decisions rather than just gut-level intuition. Going beyond opinions and biases in decision making.

Declaration of commitment   Using a formalized process to surface the depth of belief in implementing total quality service.

Delighting customers   Surprising customers with service quality beyond their expectations.

Deployment   A term used frequently in the Baldrige Award criteria to mean how thoroughly and how well the quality processes permeate the organization. Consistency is one of the keys.

Diagonal slice of the organization   Representatives that cut across both vertical layers and horizontal groups. Diagonal slices may also incorporate different units from different cities. The intent is that the group is made up of representatives from the whole organization.

Employee education and training plan   A plan that details who, when, and how the various levels of the organization will be trained in service quality.

Employee satisfaction improvement plan   A plan that determines what can be done to close the gap between employee satisfaction goals and actual satisfaction levels.

Employee well-being   Includes such issues as employee satisfaction, benefits, recognition, training, and support services (for example, recreation facilities, counseling, and day care).

Executive retreat   Meetings of the executive level that may focus on such issues as financial, strategic, or quality objectives and plans. Many of these retreats are held off-site and are facilitated by a professional to optimize executive time.

External customers   Customers who buy the service or product.

Facilitator   A person specially trained to help project teams or groups carry out their mission and goals.

Fishbone or cause-and-effect diagram   Ishikawa's tool for listing causes of problems.

Flowchart   A graphic means of depicting the steps in a process.

Focus groups   A small group led by a trained facilitator assembled for the purpose of exploring a topic or set of questions. Focus groups usually help companies explore in-depth customer needs and preferences.

Focused interviews   Same as a focus group only done one-on-one.

Gap analysis   Comparing existing reality against goals.

Hard measures of customer satisfaction   Measures that relate to actual buying behavior. Market share, revenues, and profit margins are all hard measures.

Industry trend analysis   Trends that are taking place in the whole industry. This is important in service because the bar keeps rising on customer expectations and needs. What delights customers one day is an expectation the next.

Internal audit   An audit done by company insiders.

Internal customers   Those who receive the services of others in the company. A hospital has doctors who rely on the internal services of nurses and administrators.

ISO 9000 series   A series of standards generated by the International Organization for Standardization (ISO). ISO is a worldwide federation of national standards bodies.

Joint planning   A planning process that includes the company, suppliers, and customers.

Kaizen   Continuous improvement in Japanese.

Key processes   The most critical processes to customer satisfaction and the survival of the organization.

Key quality indicators   Areas that customers have indicated are vital to their satisfaction.

Key quality objectives   Goals that pertain to key quality indicators.

Operational results   Results that relate to the operation of the company. Those results may be financial or quality measures.

Paradigm shift   A fundamental change in the way one views the world.

Pareto principle   In any phenomenon, only a few of the contributors account for the bulk of the effect.

Perceptual measure of satisfaction   Surveys that measure a customer's perception of the service or product. Those measures may include surveys, focus groups, and/or observation.

Policy   A guide to management and employee action.

Procedures manual   A manual that details the steps to be done in one or several jobs.

Process   A systematic series of steps aimed at achieving a goal.

Process capability   The inherent ability of a process to perform under current operating capabilities. The time from order to delivery may be limited by the computer systems and staffing level. Thus, the current process may only have a capability of responding within x days.

Process design   Activities used to stipulate the steps and means to achieve a process goal.

Process management   Planning and monitoring a process.

Project charter   The mission, list of activities, and schedule to be used by a quality improvement team.

Project plan for quality   A plan that spells out the infrastructure for quality; for example, the steering committee's mission, goals, and duties. Embedded in this plan are separate plans for training, accountability, visibility, celebration, and suggestions.

Project team charter   An agreement between the quality improvement team and the management about its mission, resources, and decision authority.

Quality accountability system   A methodology that incorporates quality goals in hiring, promotion, performance appraisals, and compensation.

Quality assessment system   The improvement process itself usually requires refinement. This system is designed to improve that process at regular intervals.

Quality assurance   An independent evaluation of quality-related performance. Usually it is done for those who have a need to know rather than for those directly involved.

Quality control   A process that evaluates actual performance against goals and that takes action on the difference.

Quality function deployment (QFD)   A quality tool used to translate customer needs into service features. Frequently it entails a look at how competitors are doing in those features.

Quality improvement team (QIT)   Also known as a quality circle, the QIT's purpose is to improve some facet of the service, product, or process.

Quality promoter   The Japanese term for quality champion.

Quality readiness   Captures the cultural readiness of an organization to implement TQS. Managers who have a hard time delegating authority, who consider financial goals exclusively, who resist change, who don't like to hear bad news, or who punish risk taking make it difficult for TQS to thrive.

Reengineering   Designing the service or product starting from ground zero with the customers' preferences in mind.

Root cause analysis   A structured process for determining what caused a problem. The fishbone diagram is sometimes used to help brainstorm possible causes of problems.

Scientific   Systematic observation and study.

Service   Work performed for others. Support services within an organization include such functions as word processing, payroll, expediting, engineering, maintenance, hiring, and training.

Service design (or product design)   Consciously designing the service to meet the needs of customers at the lowest cost.

Soft measures of customer satisfaction   Includes perceptual measures like customer surveys and focus groups. Soft measures don't automatically relate to hard measures of true buying behavior.

Sponsor   Same as a champion.

State of the organization   A report that provides a snapshot of the current organization. Usually it includes a section on the financial, quality, human resource, community, and competitive status of the company. It is used to do a gap analysis so that the strategic quality plan is a basis in fact from which to operate.

Statistical process control (SPC)   A term used in the 1970s through 1990s to describe the concept of using statistical tools to monitor and improve quality.

Statistical quality control (SQC)   A term used in the 1950s and 1960s to mean the same thing as SPC.

Strategic analysis   Analyzing the organization's target customers and competitive edge.

Strategic planning   A plan that uses the strategic analysis to develop goals, and an implementation plan to meet those goals.

Strategic quality plan   An integrated plan that looks at target customers, competition, and the company's strengths and weaknesses. Goals and an implementation process are usually part of the plan.

Success story system   A systematic way to deploy success stories to management, employees, suppliers, customers, shareholders, and/or the community.

Suggestion system   A systematic way to solicit ideas from employees. The means can be passive (suggestion boxes, e-mail) or active (meetings, focused interviews).

Supplier quality management   Managing the quality of the services or products of suppliers. Most high-performing TQS companies both extend their TQS training to suppliers and have rigorous measurement systems to assess their performance.

Support services quality management   Managing the quality of the support services within an organization. Support services include procurement, invoicing, legal, patents, cost and scheduling, catering, payroll, and so on.

System   A set of processes with a purpose.

Systematic   Done at regular intervals and with a complete feedback loop. Doing customer satisfaction research systematically means doing it at regular intervals and using the feedback to redesign services, policies, and procedures.

Top-down improvement   Improvements that emanate from the strategic quality planning process. Hopefully, customer needs and satisfaction are integral parts of that planning process. Each department, unit, or cross-functional team then sets goals that will help achieve the overall key quality objective.

Total quality service (TQS)   Customer satisfaction. It usually means reengineering services to met the needs of the customer, including competitive pricing needs.

TQS road map   A visual portrayal of how service quality unfolds in an organization. The implementation is actually much more holistic than the map implies.

TQS steering committee   A group that guides and coordinates the quality efforts within an organization. Unlike the executive committee, the TQS steering committee is frequently made up of a diagonal slice of the organization. Also known as the quality council.

Upside-down review   The employees rate the managers in performance review.

Visibility plan   A conscious effort to make improvement results visible to employees, suppliers, shareholders, and customers. The plan usually includes how to communicate success stories via newsletters, press releases, meetings, or electronic means.

Index

accountability. *See also* quality
     accountability systems
  negative, 45
  positive, 45
ambiguity, in customer
     satisfaction surveys, 64
American Airlines, 60
American Society for Quality
     Control (ASQC), 143
Ames Rubber, 21
annual quality reports, 47–48,
     143
approach, to total quality service,
     8–9, 143
Arendt, Carl, 78
ASQC (American Society for
     Quality Control), 143
AT&T
  benchmarking department, 78
  customer value added (CVA),
     67
AT&T Consumer Communi-
     cations Systems, 21
  Baldrige Award, 5
  customer satisfaction, 76
  performance appraisals, 46
  preparation for quality
     improvement, 12–13

AT&T Universal Card Services,
     12, 21
  delighting customers, 103
  employee empowerment, 103
  implementation of TQS, 29
  interviewing applicants, 46
  leadership in, 33
  visibility program for results,
     41
audits, 125–26
  internal, 126, 146
awards, 43, 44–45
  monetary, 108
  in suggestion systems, 108–9

Bain & Company, 3, 81
Baldrige Award, 6, 143
  business results, 129–30
  criteria, 7–8, 11
  examination process, 8–9
  and financial success, 6
  human resource development
     and management, 103, 106,
     117–18
  information and analysis
     criteria, 49–51
  leadership criteria, 31
banks, customer retention, 4

Bemowski, Karen, 12
benchmarking, 77–80, 143
  benefits, 79
  formal, 78
  informal, 78
  management sponsors, 80
  neutral services, 78
  objectives, 62
  planning, 80
  reciprocal, 79
  results, 80
  steps, 79–80
  teams, 79–80
Ben and Jerry's, 85
bias, in customer satisfaction
    surveys, 63
Bloomingdales, 84
bonus systems, 47
bottom-up employee
    involvement, 106
bottom-up quality improvement,
    10–11, 143
brainstorming, 53–54, 143
Brown, Mark Graham,
    51
Buffums, 84
Bullocks, 84
business plans, 21
business results, 129–30
  outline, 26

California Society of Quality and
    Service, 47
Camp, Robert, 77, 78
Carder, Brooks, 40
Carlson, Marilyn, 50, 76
Carlzon, Jan, 3, 104
cascade training, 33
cause-and-effect diagrams,
    53–54, 146

celebration systems, 42–45, 144
  awards, 43, 44–45
  quality fairs, 40, 43
champions, 33, 144
  role in employee involvement,
    104, 105
  and success stories, 40
  and visibility plans, 40–41
change, resistance to, 32
charters, for quality improvement
    teams, 111–12, 147
charts
  control, 50, 52–53, 62, 144
  flow-, 52, 146
  in Japanese companies, 50
  Pareto, 50, 56–57
checklists, 144
check sheets, 55–56, 144
Clark, James, 40
cluster analysis, 144
commission systems, 47
compensation systems, 47
competitors
  analysis, 144
  benchmarking, 77–80, 89
  profiles, 84–85, 89–91
complaints
  of customers, 74, 75, 144
  tracking, 75, 144
concurrent engineering, 120
consensus, growth of, 31–32
consultants, 59, 81, 106, 116
continuous improvement
    (kaizen), 32
continuous process improve-
    ment (CPI), 144
continuous quality improvement
    (CQI), 144
control charts, 50, 52–53, 62,
    144

cost of quality, 144
CPI (continuous process improvement), 144
CQI (continuous quality improvement), 144
criminals, looking for, 10–11
critical mass, for quality, 31–32
critical processes, 81, 144
cross-functional teams, 106, 144
culture, 144–45
  fun in, 6–7, 85
  total quality service, 32–33, 34, 106, 145
customer advisory groups, 75
customer appreciation days, 75
customer focus and satisfaction, 8, 59, 145
  benchmarking, 62, 77–80
  customer needs assessment, 59–60
  expectations, 60
  feedback, 74–75
  and financial success, 81
  focus groups, 62
  hard measures, 61, 146
  internal customers, 76–77
  measurement, 59, 60–61, 74
  objectives, 61, 62
  outline, 23
  soft measures, 60–61, 148
  surveys, 61–69
  over time, 62, 63
  tools, 60, 61, 62, 126
  use of data, 76
customer needs assessment, 59–60
customers. *See also* focus groups and interviews
  categories, 63
  competing, 72–74
  of competitors, 70
  complaints, 74, 75, 144
  delighting, 103, 145
  external, 146
  internal, 76–77, 146
  observing, 75
  perceptions, 61, 67
  recruiting, 4
  requirements, 120, 121
  retention, 3–4
  speakers at internal meetings, 75
customer satisfaction surveys, 61–69, 145
  ambiguity in, 64
  bias in, 63
  frequency, 64
  increasing response to, 68
  and key quality objectives, 69
  mail, 68
  objectives, 62, 63
  overall satisfaction level, 67
  questions, 63–65, 69
  representative data, 63
  results, 67, 68
  sample, 65–67
  steps, 65
  telephone, 64, 75
  testing, 67
customer value added (CVA), 67, 145

data
  comparative, 51
  for decision-making, 4–5, 49–51, 145
  driven, 4–5, 49–51, 145
  graphic form, 50
  levels, 51
  relevance, 49–50

data—*continued*
  reliability, 50
  representative, 50, 63
  trends, 51
  variability, 51
  visibility, 50
declarations of commitment
      (DOC), 36–39, 145
delighting customers, 103, 145
Deming, W. Edwards, 31, 32,
      42, 46, 98
Deming Prize, 6, 41
deployment, 8–9, 145
design review and control, 120
diagonal slices of organizations,
      35, 145
Digital Equipment, 78, 84
Disneyland, 100
documentation, of quality
      assessments, 126–27
DuPont, 78

Eastman Chemical, 21, 63, 94
Eastman Kodak, 78
education. *See* training
Electronic Data Systems, 77
employee involvement, 104–7
  bottom-up, 106
  champions' role, 104, 105
  and managers, 106–7
  motivations for, 106–7
  results, 116–17
  suggestion systems, 107–9,
      149
  top-down, 105–6
employee-of-the-month
      systems, 44
employees
  compensation, 47
  empowerment, 10–11, 32,
      116–17

interviewing, 46
  performance appraisals, 46
  ranking, 46
  recognition by, 44, 106
  recruiting, 45–46
  satisfaction, 117–18, 145
  training, 10, 104, 115–16, 145
  well-being of, 117–18, 145
  Yellow Pages systems, 86
empowerment, of employees,
      10–11, 32, 116–17
European Union, and ISO 9000
      standards, 9
executive retreats, 12, 145–46
expectations, of customers, 60
external customers, 146

facilitators, 33, 146
  in focus groups, 70, 72
  for quality improvement
      teams, 110
  for steering committees, 36
Federal Express, 11, 33, 77
feedback, customer, 74–75
financial reporting, 129–30
financial success
  and Baldrige Award, 6
  and customer satisfaction, 81
fishbone diagrams, 53–54, 146
fix-it feedback mechanisms, 74
flowcharts, 52, 146
Fluor Daniel
  benchmarking, 78
  competitor analysis, 84–85
  employee recognition, 45
  mission statement, 85
focus groups and interviews,
      70–74, 146
  benefits, 70–71
  competing customers in,
      72–74

focus groups and interviews—
    continued
    customers of competitors,
        70
    facilitators, 70, 72
    follow-up, 72
    incentives, 71
    invitations, 70, 71
    and key processes, 95
    location, 71
    objectives, 62
    questions, 72, 73
    taping, 72
Ford Motor Company, 109
Fox Valley College, 43
Fuji Xerox, 21
fun, 6–7, 85

Gallery Furniture Store,
    47
gap analysis, 13, 146
Gates, Bill, 85
General Accounting Office,
    6
General Motors, Saturn, 12, 47,
    77
goals
    in declarations of
        commitment, 38
    for individuals, 46
    long-term, 82
    negotiating, 48
government regulation, 84
graphs. *See* charts
Greenleaf, Robert, 13
GTE Directories
    Baldrige Award, 5
    customer focus, 8, 49–50,
        76
    planning, 21
gut-level management, 49

Haas, Robert, 14
hard measures
    of customer satisfaction, 61,
        146
    of employee satisfaction, 118
    key quality indicators, 97, 98
Hewlett-Packard, 21, 74, 82
hiring process, 45–46
Holiday Hotels, 106
hospitals, 84
house of quality, 121
human resource development
        and management
    Baldrige Award criteria, 103,
        106, 117–18
    compensation systems, 47
    employee involvement, 104–7
    empowerment, 10–11, 32,
        116–17
    hiring, 45–46
    measures, 117–18
    outline, 25–26
    performance appraisals, 46
    quality improvement teams
        (QITs), 110–14, 148
    in strategic plan, 104
    suggestion systems, 107–9,
        149
    training, 10, 104, 115–16
human resource plans, 104, 105

IBM, 40, 78, 84, 108
industry trend analysis, 84, 89,
    146
Indy 500 pit crews, 77
Infiniti, 47
information and analysis
    Baldrige Award criteria, 49–51
    check sheets, 55–56, 144
    control charts, 50, 52–53, 62,
        144

information and analysis—
 continued
 fishbone diagrams, 53–54
 importance of, 49
 outline, 22–23
 Pareto charts, 50, 56–57
 tools, 51–52
 use of tools, 57–58
inspection, 98–100
internal audits, 126, 146
internal customers, 76–77, 146
interviews
 focus, 70–74
 of job applicants, 46
ISO (International Organization
 for Standardization), 9
ISO 9000 series standards, 9–10,
 146
 procedures, 100–102
 quality control plans, 99, 100

Japan
 annual quality reports, 47, 48
 champions, 33
 Deming Prize, 6
 kaizen, 32, 146
 kaizen corners, 50
 management involvement in
  quality, 11, 31–32
 quality conferences, 43
 quality improvement teams,
  110
 service companies, 110
 strategic planning, 110
 suggestion systems, 107, 110
 total quality in, 6
 training, 104
Japanese Union of Scientists
 and Engineers (JUSE), 4, 32
J.D. Powers, 78

John Deere, 77
joint planning, 125, 146
Juki, 21, 41, 50, 83
Juran, Joseph M., 56
JUSE (Japanese Union of
 Scientists and Engineers), 4,
 32

Kahn, Paul, 12
kaizen, 32, 146
 corners, 50
Kano, Noriaki, 4, 10, 31, 49, 107
Kearns, David, 40
Kelleher, Herb, 7
Kelly, Tricia, 107
key processes, 24, 93–96, 146
 improving, 121–23
key quality indicators, 96, 97, 146
 communicating to suppliers,
  125
 hard measures, 97, 98
 and key processes, 95, 121
 soft measures, 97, 98
key quality objectives (KQO),
 96–97, 146
 from customer satisfaction
  surveys, 69
 levels, 97, 98
 and training, 115
key selling points, 92
KKD management, 49
Komatsu, 41, 50, 52, 83

leadership
 Baldrige Award criteria, 31
 outline, 21–22
 training, 13–14, 32–33
 unresponsive, 14
 value-based, 13
 virtual, 13–14

Lerner, 84
Levi Strauss, 14
Lexus, 47
Lieb, George, 21
Likert scale, 65
The Limited, 84
L.L. Bean, 77
looking for criminals or lovers,
     10–11
lovers, looking for, 10–11
lower control limits (LCL), 52, 53

mail surveys, 68
malls, marketing, 84
management. *See also* leadership
   authoritarian, 107
   and benchmarking, 80
   coaching by, 106
   and employee involvement,
      106–7
   gut-level, 49
   middle, 32, 116
   phone surveys by, 75
   resistance to change, 32
   role in quality improvement,
      11, 31–32
   training, 116
Manufacturer's Alliance for
     Productivity and Innovation
     (MAPI), 116–17
Marriott, 33, 75, 77, 100
MBNA, 74
McClellan, Stephen T., 77
McIngvale, James, 47
Michael, Buck, 45
Microsoft, 77, 85
middle management, 32, 116
Milliken
   benchmarking of, 77
   leadership in, 33

skill-based compensation, 47
   suggestion system, 108
mission statements, 83, 85
monetary awards, 108
motivation. *See* celebration
     systems; spirit
Motorola, 21, 33, 82

Nacchio, Joe, 76
National Rifle Association, 63
negative accountability, 45
new product design systems,
     119–20, 148
Nippon Denso, 21, 41
Nissan, 41, 50
North Federal Credit Union of
     San Diego, 47

observation, of customers, 75
operational results, 129–30, 146
outsourcing, 77
overall satisfaction level, 67

paradigm shifts, 33, 34, 147
Pareto, Vilfredo, 56
Pareto charts, 50, 56–57
Pareto principle, 56, 147
pep rallies, 85
perceptions
   of customers, 61, 67
   and key quality objectives, 97,
      98
perceptual measures of satisfac-
     tion, 98, 147
performance appraisals, 46
Perot, Ross, 77
Peters, Tom, 11
policies, 147
positive accountability, 45
positive climate, 6–7, 85

Prism Radio Partners, 115
procedures manuals, 100–102, 147
processes, 93–94, 147
    capabilities, 147
    controlling, 122–23
    critical, 81, 144
    design, 147, 148
    improving, 121–23
    key, 24, 93–96, 121
process management, 119,
    147
    new product design systems,
        119–20, 148
    outline, 26
    production and delivery
        quality management,
        121–23
    quality assessment, 125–27
    quality function deployment
        (QFD), 121, 122, 148
    supplier quality management,
        124–25
    support services quality sys-
        tems, 123–24, 149
product design systems, 119–20,
    148
production and delivery quality
    management, 121–23
profits, and customer retention,
    4
project charters, for quality
    improvement teams,
    111–12, 147
project methodology, for quality
    improvement teams,
    113–14
project plans for quality, 36, 37,
    147
purchasing function, 125

QIT. See quality improvement
    teams
quality accountability systems,
    45–47, 147
    compensation, 47
    hiring, 45–46
    performance appraisals, 46
quality analysis, 82, 91–93
    key processes, 24, 93–96, 146
    key quality indicators, 96, 97,
        146
    key quality objectives (KQO),
        96–97, 98, 146
    steps, 93
quality assessment systems,
    125–27, 147
    documentation, 126–27
quality assurance, 147
quality audits, 125–26
quality books, 83
quality circles. See quality
    improvement teams
quality control, 147
    plans, 98–100
    statistical, 148
quality fairs, 40, 43
quality function deployment
    (QFD), 121, 122, 148
quality improvement
    bottom-up, 10–11, 143
    failures, 107
    looking for criminals or
        lovers, 10–11
    top-down, 11, 149
quality improvement teams
    (QITs), 110–14, 148
    documentation, 127
    facilitators, 110
    forming, 110–11

quality improvement teams
   (QITs)—*continued*
   in Japan, 110
   meetings, 111
   members, 111
   project charters, 111–12, 147
   project methodology, 113–14
   schedules, 114
   scribes, 111
   themes for, 106
   tools, 113
quality plans, 81, 82
quality promoters, 33, 148
quality readiness, 12–14, 148
questionnaires, 50
   customer satisfaction, 63–65, 69
   Likert scale, 65
   state-of-the-organization
      report, 87–89

Rancho Vista Bank, 44
readiness, for quality
      improvement, 12–14, 148
recognition programs, 40, 106
   celebration systems, 42–45,
      144
   informal, 44
   in suggestion systems, 108–9
recruiting, quality orientation,
      45–46
reengineering, 148
reference checks, 45–46
Reichheld, Frederick F., 3–4
relevance, of data collected,
      49–50
reliability, of data collected, 50
reporting, business results,
      129–30
representative data, 50, 63

results
   of benchmarking, 80
   business, 129–30
   of total quality service, 8–9
   visibility plans, 40–42, 150
rewards. *See* recognition
      programs
Ritz-Carlton, 11, 21, 74
road map, total quality service,
      19–21, 149
Robinsons, 84
root cause analysis, 54, 148

sales staff, compensation, 47
Saturn, 12, 47, 77
Scandinavian Airlines System, 3,
      104
scientific analysis, 4–5, 148
self-inspection, 100
services, 148
   design of, 148
   support, 123–24, 149
skill-based compensation, 47
soft measures
   of customer satisfaction,
      60–61, 148
   of employee satisfaction, 118
   key quality indicators, 97, 98
Solectron, 21
Southwest Airlines, 6–7, 13, 60,
      77
Spa Resort Hawaiian, 21, 41, 75
SPC (statistical process control),
      51, 148
Spendolini, Michael J., 79
spirit, 106–7
sponsors. *See* champions
state-of-the-organization
      reports, 82, 86–89, 148

statistical process control (SPC),
    51, 148
statistical quality control (SQC),
    148
steering committees, 34–36, 150
    actions, 34–35
    and celebrations, 42
    declarations of commitment
        (DOC), 36–39
    and employee involvement,
        105–6
    meetings, 36
    membership, 35
    project plans, 37
    team building in, 36
    training, 81, 116
strategic analysis, 82, 83–86, 149
    competitor profiles, 84–85,
        89–91
    industry trend analysis, 84, 89,
        146
    key selling points, 92
    mission statements, 85
    responsibility for, 85–86
    synthesis, 85, 91, 92
strategic planning, 149
    components, 82
    consultants for, 81
    in Japan, 110
    outline, 23–25
    process, 82
strategic plans, 81, 82, 97–98,
    149
    compared to business plans,
        21
    compared to quality control
        plans, 99
    contents, 99
    frequency, 82
    human resource planning, 104
    size, 83

success story systems, 39–40, 149
    benefits, 39
    implementation, 39–40, 127
suggestion systems, 107–9, 149
    forms, 107, 108
    recognition, 108–9
    review systems, 107–8
    testing, 109
supplier quality management,
    124–25, 149
support services quality manage-
    ment, 123–24, 149
surveys. See customer
    satisfaction surveys
systematic programs, 5, 149
systems, 93, 149

Tague, Nancy, 52
teams
    benchmarking, 79–80
    building in steering
        committees, 36
    cross-functional, 106, 144
teamwide quality programs, 5
Technical Assistance Research
    Program Institute, 123
telephone surveys, 64, 75
360° reviews, 46, 143
top-down employee involve-
    ment, 105–6
top-down quality improvement,
    11, 149
Torki, Nevienne, 52
Toshiba, 76
total quality service (TQS), 3–6,
    149
    components, 4–5
    critical processes, 81
    culture, 32–33, 34, 106, 145
    implementation schedules, 14,
        15, 29

total quality service (TQS)—
    continued
    outline, 21–26
    readiness, 12–14, 148
    road map, 19–21, 149
Toyo Engineering, 41
Toyota, 50, 83, 107
TQS. *See* total quality service
tracking systems, 75, 127, 144
training, 10, 115–16
    by consultants, 81, 116
    cascade, 33
    of frontline workers, 104, 116
    in Japan, 104
    and key quality objectives,
        115
    leadership, 13–14
    management, 116
    plans, 25, 115–16, 145
    of steering committees, 81,
        116
    topics, 116
transaction feedback forms, 75
trends, 51, 129
    industry, 84, 89, 146
Trish, 78

upper control limits (UCL), 52,
    53
upside-down reviews, 12, 150

value-based leadership, 13
values, 84
variability, of data, 51

variations, in control charts, 52
Victoria's Secret, 84
virtual leadership, 13–14
visibility plans for results, 40–42,
    150
vision statements, 85

Wainwright, 21
    Baldrige Award, 5
    benchmarking, 78
    customer surveys, 76
    profit sharing, 47
Wal Mart, 85
Western Electric, 98
Westinghouse Productivity and
        Quality Center, 78

Xerox, 21
    benchmarking, 77
    Teamwork days, 40, 43
    training, 33

Yellow Pages systems, 86
Y Hewlett-Packard, 21

Zytec, 21
    Baldrige award, 6
    benchmarking, 78
    cross-functional teams, 105–6
    daily quality meetings, 76
    employee empowerment, 11,
        106
    employee recognition, 44, 106
    quality books, 83

About the Author

Sheila Kessler is a principal at Competitive Edge, a consulting firm that specializes in helping organizations set up service quality programs, training, and customer needs and satisfaction systems. She has worked with more than 50 Fortune 500 companies. Sheila teaches total quality service and customer satisfaction measurement at the University of Phoenix and also developed and teaches a customer satisfaction retention course for ASQC.

DATE DUE

DATE DUE			
OCT 1 5 1996			
MAR 2 1 2001			
GAYLORD			PRINTED IN U.S.A.